Architecture-Aware Optimization Strategies in Real-time Image Processing

Series Editor
Henri Maître

Architecture-Aware Optimization Strategies in Real-time Image Processing

Chao Li
Souleymane Balla-Arabe
Fan Yang-Song

iSTE

WILEY

First published 2017 in Great Britain and the United States by ISTE Ltd and John Wiley & Sons, Inc.

ISTE Ltd
27-37 St George's Road
London SW19 4EU
UK

www.iste.co.uk

John Wiley & Sons, Inc.
111 River Street
Hoboken, NJ 07030
USA

www.wiley.com

Library of Congress Control Number: 2017948985

British Library Cataloguing-in-Publication Data
A CIP record for this book is available from the British Library
ISBN 978-1-78630-094-2

MIX
Paper from responsible sources
FSC www.fsc.org FSC® C013604

Contents

Preface

In the image processing field, a lot of applications require real-time execution in the domains of medical technology, robotics and transmission, etc. Recently, real-time image processing fields have made a lot of progress. Technological developments allow engineers to integrate more complex algorithms with large data volumes onto embedded systems, and produce series of new sophisticated electronic architectures at an affordable price. At the same time, industrial and academic researchers have proposed new methods and provided new tools in order to facilitate real-time image processing realization at different levels. It is necessary to perform a deepened educational and synthetic survey on this topic. We will present electronic platforms, methods, and strategies to reach this objective.

This book consists of seven chapters ranging from the fundamental conceptual introduction of real-time image processing to future perspectives in this area. We describe hardware architectures and different optimization strategies for real-time purposes. The latter consists of a survey of software and hardware co-design tools at different levels. Two real-time applications will be presented in detail in order to illustrate the proposed approaches.

The major originalities of this book include (1) algorithm architecture mapping: we select methods and tools that treat simultaneously the application and its electronic platform in order to perform fast and optimal design space exploration (DSE), (2) each approach will be illustrated by concrete examples and (3) two of the chosen algorithms have been only recently advanced in their domain.

This book is written primarily for those who are familiar with the basics of image processing and want to implement the target image processing design using different electronic platforms for computing acceleration. It accomplishes this by presenting the techniques and approaches step by step, the algorithm and architecture conjointly, and by notions of description and example illustration. This concerns both the software engineer and the hardware engineer.

This book will also be adequate for those who are familiar with programming and applying embedded systems to other problems and are considering image processing applications. Much of the focus and many of the examples are taken from image processing applications. Sufficient detail is given to make algorithms and their implementation clear.

Chao LI
Souleymane BALLA-ARABE
Fan YANG
August 2017

1

Introduction of Real-time Image Processing

1.1. General image processing presentation

The traditional view of an image derives heavily from experience in photography, television and the like. This means that an image is a two-dimensional (2D) structure, a representation and also a structure with meaning to a visual response system. This view of an image only accepts spatial variation (a static image). In parallel, a dynamic image has spatial and temporal variation. In most contexts, this is usually referred to as video. This more complex structure needs to be viewed as a sequence of images each representing a particular instance in time. On the other hand, an image can be formed by taking a sampling plane through that volume and so the variation in three dimensions is observed. This may be referred to as a volume image. An image linked to a volume that changes with time is a further possibility. This has particular significance in medical imaging applications [MYE 09].

Image processing is a method to convert an image into digital form and perform some operations on it in order to get an improved image or to extract some useful information from it. A digital image described in a 2D space is usually considered as 2D signals while applying already set signal methods to it. Image processing forms a core research area within engineering and computer science too.

Image processing is an enabling technology for a wide range of applications including remote sensing, security, image data sets, digital television, robotics and medical imaging, etc. The goal of this technology can be divided into three levels: low, medium and high. Low-level image processing techniques are mathematical or logical operators that perform simple processing tasks. This "processing" level possesses both *image_in* and *image_out*. Medium-level image processing combines the simple low-level operators to perform feature extraction and pattern recognition functions. Using an *image_in*, this "analysis" level produces *measurements_out* (parameters). High-level image processing uses combinations of medium-level functions to perform interpretation. This "understanding" level treats *measurements_in* for high-level *description_out*. Usually, apart from these three levels, it is also necessary to apply preprocessing techniques that are designed to remove distortions introduced by sensors.

Figure 1.1. *Overview of the typical image acquisition process (see [MOE 12])*

Before any image processing can commence, an image must be captured by a camera and converted into a manageable entity. This is the image acquisition process (see Figure 1.1), which consists of three steps; energy reflected from the object of interest, an optical system that focuses on the energy and finally a sensor that measures the amount of energy. In order to capture an image, a camera requires some sort of measurable energy. The energy of interest in this context is light or electromagnetic waves. Each wave can have different wavelengths (or different energy levels or different frequencies). The human visual spectrum is in the range of approximately 400–700 nm.

After having illuminated the object of interest, the light reflected from the object now has to be captured by the camera composed of an optical system and an image sensor. The role of the first is to focus the light reflected from the object onto the second (a material sensitive to the reflected light). An image sensor consists of a 2D array of cells. Each of these cells is denoted a pixel and is capable of measuring the amount of incident light and convert that into a voltage, which in turn is converted into a digital number. The more incident light, the higher the voltage and the higher the digital number.

In order to transform the information from the sensor into an image, each cell content is now converted into a pixel value in the range: (0, 255). Such a value is interpreted as the amount of light hitting a cell. This is denoted the intensity of a pixel. It is visualized as a shade of gray denoted gray-level value ranging from black (0) to white (255). Standardly, a monochromatic, static image corresponds to a matrix of m rows and n columns. Therefore, the camera records $m \times n$ pixels of 8 bit values.

In order to capture a color image, the color camera must record three matrices of three primary colors red, green and blue. Recently, a lot of applications are realized using multispectral image processing, since the multispectral cameras are more available with reasonable prices. According to application needs, multispectral images are captured using different wavelengths (bands). They can be considered as a cube formed by 2D gray-level images. Figure 1.2 displays two typical test images in the area of image processing research. Figure 1.3 gives two multispectral cube examples; the right image is captured for a skin lesion assessment application.

Figure 1.2. *Lena (gray-level image) and landscape (color image). For a color version of the figure, see www.iste.co.uk/li/image.zip*

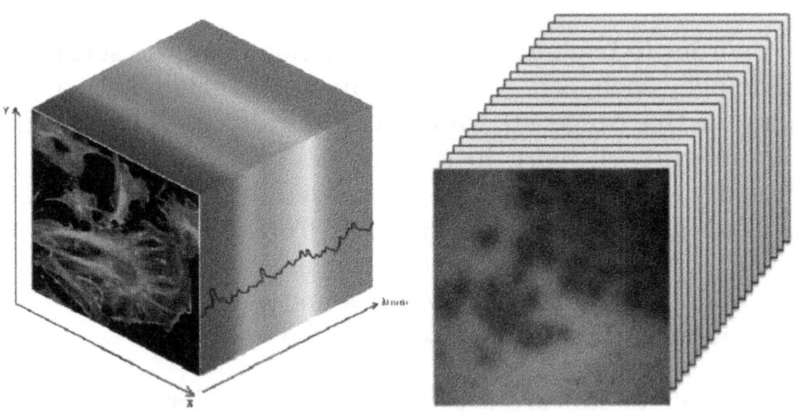

Figure 1.3. *Two multispectral image cubes. For a color version of the figure, see www.iste.co.uk/li/image.zip*

Certain tools are central to the processing of digital images. These include mathematical tools, statistical descriptions and manipulative tools. We can cite some more used such as *convolution*, *filtering in spatial and frequency domains*, *morphological operations* and *image transforms*, etc. The types of basic operations that can be applied to digital images to transform an input image A(m, n) into an output image B(m, n) (or another representation) can be classified into three categories:

– Point: the output value at a specific matrix coordinate is dependent only on the input value at that same coordinate. In this case, generic operation complexity per pixel is constant.

– Local: the output value at a specific coordinate is dependent on the input values in the *neighborhood* of that same coordinate. In most cases, the type of neighborhood is rectangular with 8-connected (3 × 3 mask) or 24-connected (5 × 5 mask pixels). The complexity per pixel is proportional to the square of neighborhood size.

– Global: the output value at a specific coordinate is dependent on all the values in the input image. The complexity per pixel is equal to $N \times N$ (image size = N).

The complexity per pixel of each operation type participates in the total complexity of an image processing chain for a target application. This total complexity per image decides the processing speed (time performance).

1.2. Real-time image processing

Real-time image processing is the subfield of image processing focused on producing and analyzing images in real time. Generally, a real-time system has the following three interpretations within different senses [KEH 06]:

– real time in the perceptual sense, which is used to describe the interaction between a human and a computer device for a near instantaneous response of the device to an input by a human user;

– real time in the software engineering sense, which is used to describe a concept of bounded response time in the computer device. With this constraint, the device must satisfy both the correctness of the outputs and their timeliness;

– real time in the signal processing sense, which is used to describe the constraint within which the computer device has to complete processing in the time available between successive input samples.

Since image processing is a subfield of signal processing, in this book we base the interpretation of “real time” on the signal processing sense.

In order to satisfy the constraint of “real time”, Kehtarnavaz [KEH 11] points out that the total instruction count of a real-time algorithm must be “less than the number of instructions that can be executed between two consecutive samples”, and Ackenhusen *et al.* [ACK 99] describe the “real-time processing” as a computation of “a certain number of operations upon a required amount of input data within a specified interval of time” as well. Therefore, the key technique of real-time image processing is to ensure that the amount of time for completing all the requisite transferring and processing of image data is less than the allotted time for processing. For example, if the algorithm is aimed at an entire frame (a static image) and the frame frequency of the system is 30 frames per second (fps), the processing of a single frame should be finished during 33 ms.

Real-time image processing systems involve processing vast amounts of image data in a timely manner for the purpose of extracting useful information, which could mean anything from obtaining an enhanced image to intelligent scene analysis. Digital images and video are essentially multidimensional signals and are thus quite data intensive, requiring a significant amount of computation and memory resources for their processing. A common theme in real-time image processing systems is how to deal with their vast amount of processing and computations. The key to cope with this issue is the concept of parallel processing, a concept well known to those working in the architecture area who deal with computations on large data sets [KEH 06]. In fact, much of what goes into implementing an efficient image processing system centers on how well the implementation, in both hardware and software, exploits different forms of parallelism in an algorithm, which can be data level parallelism (DLP) and/or instruction level parallelism (ILP). DLP manifests itself in the application of the same operation on different sets of data, while ILP manifests itself in scheduling the simultaneous execution of multiple independent operations in a pipeline fashion.

Low-level image processing transforms image data to another image data. This means that such operators deal directly with image matrix data at the pixel level. Examples of such operations include color transformations, linear or nonlinear filtering, noise reduction and frequency domain transformations. In this low-level processing, one can observe three operation categories. Point operations are the simplest since a given input pixel is transformed into an output pixel, where the transformation does not depend on any of the pixels surrounding the input pixel. Local neighborhood operations are more complex in that the transformation from an input pixel to an output pixel depends on a neighborhood of the input pixel. Such operations include 2D spatial convolution and filtering, smoothing, sharpening, etc. These operations require a large amount of computations. Finally, global operations build upon neighborhood operations in which a single output pixel depends on every pixel in the input image. An example of such an operation is the discrete Fourier transform that depends on the entire image. These operations are quite data intensive as well.

All low-level image operations involve nested looping through all the pixels in an input image with the innermost loop applying a point, neighborhood or global operator to the pixels forming an output image. Therefore, these are fairly data-intensive operations, with highly structured and predictable processing, requiring a high bandwidth for accessing image data. In general, low-level operations are excellent candidates for exploiting DLP.

Medium-level operations transform image data to a slightly more abstract form of information by extracting certain features from an image. This means that such operations also deal with the image at the pixel level for input, but the transformations involved cause a reduction in the amount of data from input to output. Medium-level operations primarily include segmenting an image into regions/objects of interest or extracting edges, lines, or other image attributes such as statistical features. The goal these operations is to reduce the amount of data to form a set of features suitable for further high-level processing [KEH 06]. Some medium-level operations are also data intensive with a regular processing structure, thus making them suitable candidates for exploiting DLP.

High-level operations interpret the abstract data from the medium level, performing high-level knowledge-based scene analysis on a reduced amount of data. Such operations include classification/recognition of objects or a control decision based on some extracted features. These types of operations are usually characterized by control or branch-intensive operations. Thus, they are less data intensive and more inherently sequential rather than parallel. Due to their irregular structure and low bandwidth requirements, such operations are suitable candidates for exploiting ILP, although their data-intensive portions usually include some form of matrix-vector operations that are suitable for exploiting DLP.

From the above discussion, one can see that there is a wide range of diversity in image processing. A typical image processing chain combines the three levels of operations into a complete system, as shown in Figure 1.4, where top shows the image processing chain and bottom shows the decrease in the amount of data from the start of the chain to the end for an N × N image with P bits of precision. The diversity of operations in image processing leads to the understanding that a single processor might not be suitable for a real-time image processing algorithm implementation. A more appropriate solution would thus couple multiple computation components of different characteristics [KEH 06].

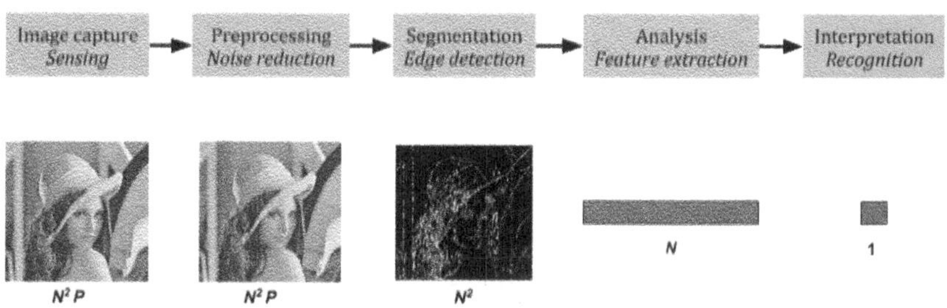

Figure 1.4. *Diversity of operations in image processing: typical processing chain (top) and decrease in amount of data across processing chain (bottom)*

Bearing in mind the above argument, developing a real-time processing system can be quite a challenge. The solution often ends up as some combination of hardware and software approaches. From the hardware point of view, the challenge is to determine what kind of hardware is best suited for a given image processing task among a myriad of available hardware platform choices. From the algorithm and/or software point of view, the challenge involves being able to guarantee that "real-time" deadlines are met, which could involve making choices between different algorithms based on computational complexity, algorithm optimization in the parallel execution sense, using a real-time operating system, and extracting accurate timing measurements from the entire system by profiling the developed algorithm [KEH 06].

In the real-time image processing area, embedded systems are often involved. A precise definition of an embedded system is not easy. Simply stated, all computing systems other than general purpose computers (with monitor, keyboard, etc.) are embedded systems. An embedded system has software embedded into hardware, which makes a system dedicated to an application or a specific part of an application. It is an engineering artifact involving computation that is subject to physical constraints (reaction constraints and execution constraints) arising through interactions of computational processes with the physical world. Embedded image processing systems are typically designed to meet real-time constraints; a real-time system reacts to stimuli from the controlled object/operator within the time interval dictated by the environment.

When the complexity of the image processing applications leads to a high ratio between its computation volume and its reaction time, standard off-the-shelf sequential architectures are inadequate. Parallel, distributed and multicore architectures are required. Programming such architectures is an order of magnitude harder than with uniprocessor sequential ones, and even more so when architecture resources must be minimized to match cost, power and volume constraints required for embedded applications.

In this context, hardware platform selection and careful and fine-tuned application programming/development environments become more and more important. At the same time, the application market is quickly spreading which reduces the time available for the design of individual applications. These facts increase the demand of rapid prototyping of real-time image processing. The rapid prototyping of complex parallel real-time embedded applications is essentially based on the software/hardware co-design. This notion is knows in two senses; for multicomponent architecture, this software/hardware (SW/HW) co-design step distributes some parts of the applications to processors by running software, while other parts must be implemented by hardware running on specific integrated circuits. On the other hand, SW/HW co-design usually means application design using both SW/HW development environments.

Designing real-time image processing systems is a challenging task indeed. Practical issues of speed, accuracy, robustness, adaptability, flexibility and total system cost are important aspects of an embedded system design. In practice, one usually has to trade one aspect for another. Since the design parameters depend on each other, the trade-off analysis can be viewed as a system optimization problem in a multidimensional space with various constraint curves and surfaces. This design space exploration task, from a mathematical viewpoint, consists of determining optimal design working points.

Today, we are at a crossroad in the development of real-time image processing systems. The advancements in integrated circuit technology make it now feasible to put to practical use the rich theoretical results obtained by the image processing community. In spite of the fact that the value of an algorithm hinges upon the ease with which it can be placed into practical use, the implementation challenges involved have often discouraged researchers from pursuing the idea further, leaving it to hardware experts to implement a practical version in real time. The purpose of the following chapters is to facilitate this task by providing a broad overview of the advanced strategies/tools for rapid prototyping of real-time image processing system.

The rest of the book is organized as follows: Chapter 2 describes commonly used hardware architectures for real-time image processing. After a comparison of the currently available platforms and their development environments, we concentrate on rapid prototyping of image processing based on field programmable gate array (FPGA) technology. Chapter 3 presents research results about enabling the reconfigurable instruction set processor model by exploiting FPGA technology for discrete cosine transform algorithm implementation. High-level synthesis (HLS) technique is introduced in Chapter 4 with a skin lesion assessment application as an illustration example. In Chapter 5, we propose a novel source-to-source compilation strategy in order to improve HLS design performances. This CDMS4HLS technique is tested and validated by the embedded implementation of very high resolution satellite image segmentation in Chapter 6. Chapter 7 examines real-time image processing with very high level synthesis.

2

Hardware Architectures for Real-time Processing

2.1. History of image processing hardware platforms

In the 1960s, NASA Scientifics began to use digital cameras for digital image processing onto workstations with continually increasing computation power. When an image processing system required a real-time throughput, multiple boards with multiple computers working in parallel were used. In the 1980s, digital signal processors (DSPs) were created in order to accelerate the computation necessary for signal processing algorithms. DSPs helped to usher in the age of portable embedded computing.

The mid-1980s also saw the introduction of programmable logic devices such as the field programmable gate array (FPGA), a technology that aimed to unite the software flexibility through programmable logic with the speed of dedicated hardware such as application-specific integrated circuits (ASIC). In the 1990s, there was further growth in both DSP performance, through increased use of parallel processing techniques, and FPGA performance to meet the needs of multimedia devices [KEH 06]. The concept of system-on-chip (SoC) was introduced during this period; it sought to bring all necessary processing power for an entire system onto a single chip. In addition to these developments, the massive parallel computation power of graphics processing units (GPUs) is used for performing compute-intensive image processing algorithms; they are also considered to be embedded devices.

In the following sections, we briefly describe these computation devices and programming/development environments for their use. This chapter is concluded by a comparison and discussion of available hardware platforms for real-time image processing.

2.2. General-purpose processors

There are two types of general-purpose processors (GPPs) on the market today, geared towards either embedded or non-embedded applications. Embedded GPPs are often used in multicore embedded SoC providing the horsepower to cope with control- and branch-intensive instructions. Actual desktop GPPs are high-performance processors with highly parallel architectures, containing features that help to exploit ILP in control-intensive, high-level image operations. SIMD (single instruction multiple data: a processing concept) extensions have also been incorporated in their instruction sets in order to exploit DLP corresponding to low- and medium-levels image operations.

As shown in Figure 2.1, GPPs' architecture is based on the *Von Neumann* model, that is program and data share a single external memory and bus. In order to break the speed bottleneck due to the limited throughput (data transfer rate) between the processing core and memory, actual GPPs usually contain two separate caches (one for program and another for data) that connect directly to the processing core.

For the GPP-based designs, both desktop and embedded GPPs are supported by mature development tools and efficient compilers, allowing quick development cycles. Therefore, the users almost do not need to worry about any hardware constraints and can concentrate more on software development.

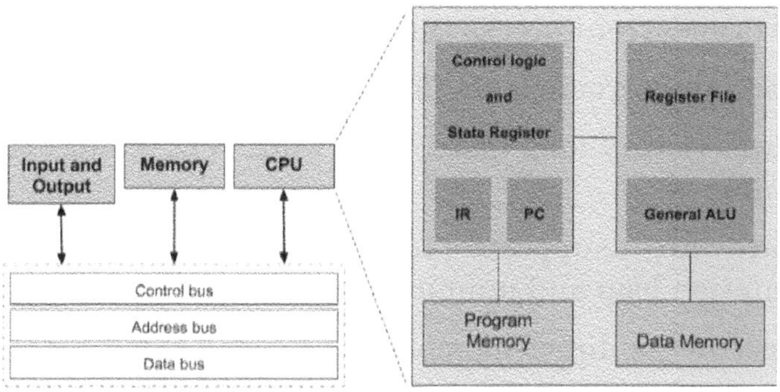

Figure 2.1. *Typical architecture of GPP (left) and GPP's processing core (right)*

2.3. Digital signal processors

A DSP is a specified microprocessor whose architecture is optimized for digital signal processing. Its goal is to meet the performance requirements of digital signal processing applications in terms of efficiency. From the 1980s to the present, the applications of DSPs have spread from the processing of low-frequency signals, i.e. speech and sonar signal processing, to today's computationally intensive image processing. Meanwhile, its architecture has also evolved from special DSP to configurable ones [KEH 06]. With enhancement of its performance–cost ratio, it is increasingly widely used in the field of real-time image processing.

The architecture of DSP is based on the *Harvard* architecture, which separates program and data memories. It usually has customized hardware multipliers, is optimized by operation pipelining and supports special digital signal processing instructions. Generally, the main features of DSPs include (see Figure 2.2):

– separate program and data memories;

– integrated high-speed random access memory and support concurrent access through separate data bus;

– special instruction set for SIMD;

– support very long instruction word (VLIW) for operation pipelining;

– hardware supports for low-consumption or non-consumption loop and jumping controls;

– execute as direct memory access device within the host environment;

– support both analog to digital converter and digital to analog converter input/output (I/O) protocols.

Up to present, the main worldwide digital signal processor manufacturers include Texas Instruments (TI), Motorola, Lucent, etc., in which TI occupies the most market share. C5x and C6x are the two categories of TI's products that are the most widely used in the world. The C5000 ultra-low-power DSP platform includes a broad portfolio of the industry's lowest power 16-bit DSPs with performance up to 300 MHz (600 MIPS); ideal for portable devices in audio, voice and vision, and other applications requiring analytics with ultra-low-power. In contrast, the C6000 category includes a number of high-performance digital signal processor families, such as C66x, the world's fastest floating-point core with devices ranging from single core C6652 to octal core C6678 and supporting core speeds up to 1.4 GHz, and C64x, a single or multicore fixed point processor with speeds up to 1.2 GHz. Additionally, in order to further add to the system performance in terms of cost, power and area savings, the normal DSPs are sometimes combined with the

Advanced RISC (reduced instruction set computing) Machines (ARM) to perform a heterogeneous platform, i.e. ARM9 + C674x and quad-core Cortex-A15 + 8xC66x cores. This technology provides the necessary processing and memory bandwidth to achieve a complete imaging SoC.

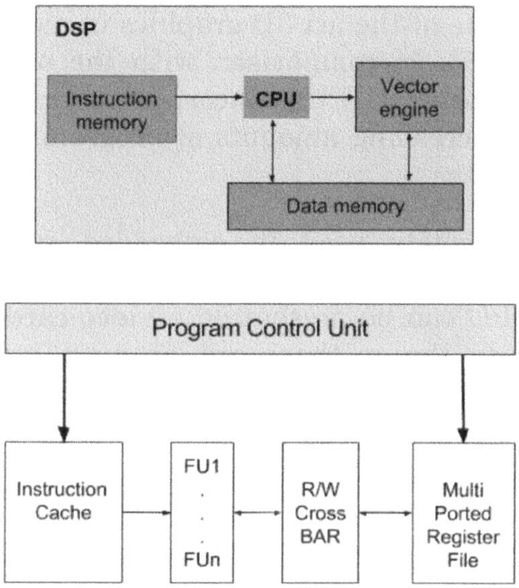

Figure 2.2. *Typical Harvard architecture and block diagram of VLIW architecture*

Since DSPs contain specific architectural features that can effectively speed up repetitive and compute-intensive signal processing routines, they become a viable option for inclusion in a real-time image processing system. Nevertheless, DSP is fully programmable, which adds to its inherent flexibility to software updates. Modern development tools such as efficient C code compilers and hardware-specific intrinsic functions have supplanted the need to generate hand-coded assembly for all but the most critical core loops. This leads to more efficient development cycles and faster time to market.

2.4. Graphics processing units

The early 2000s witnessed the introduction of this new type of processor. The primary function of GPUs is for real-time rendering of three-dimensional (3D) computer graphics enabling fast frame rates and highers level of realism required for state-of-the art 3D graphics in modern computer games [KEH 06]. In comparison with the originals GPUs, current generation GPUs incorporate more flexibility through ever-increasing amounts of programmability vertex and texture units.

GPUs are widely used in embedded systems, mobile phones, personal computers and game consoles, etc. In practice, a GPU can be present on a video card, or it can be embedded onto the motherboard or even into the CPU-integrated circuits. The architecture of modern GPUs makes them more effective than GPPs for high-parallelism algorithms (see Figure 2.3).

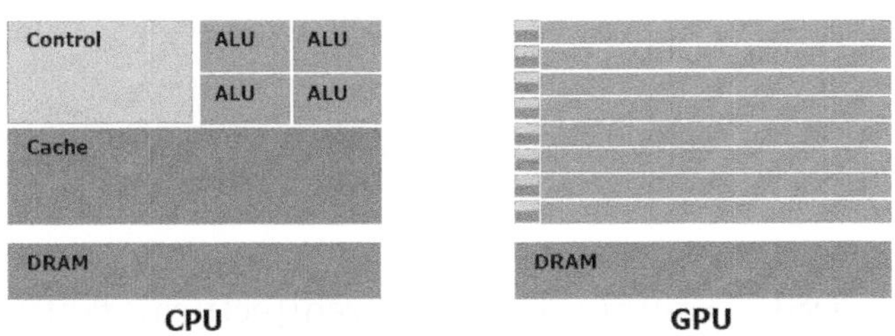

Figure 2.3. *Typical architecture of GPU in comparison with CPU. DRAM, dynamic random access memory; ALU, arithmetic logic unit*

The key technique of GPU is multiprocessors with massive parallelism. The schematic diagram shown in Figure 2.3 contains multiple microprocessors, and each of them consists of multiple processing cores. For example, NVIDIA, which is one of the most famous GPU manufacturers, delivers G80 GPU with eight multiprocessors and 128 cores in total. In each clock, each core can produce a computation result, giving this design a very high peak performance rating. Furthermore, each multiprocessor executes in SIMD mode; it means that all basic computation units execute the same instruction simultaneously.

Actually, there are variant available development environments for GPU-based designs, including Compute Unified Device Architecture (CUDA) [NVI 12] language, Open Computing Languages (OpenCL) [NVI 11], C++ Accelerated Massive Parallelism (C++ AMP) and Matrix laboratory (Matlab) [MAT 15]. Together with the high computation capabilities, the increased level of programmability and the fact that GPUs can be found in almost every desktop and portable PC today, a great number of researchers have exploited GPUs for their designs.

GPUs have already been deployed to solve real-time image processing problem. GPU-based image processing can be performed by first downloading all image segments (small part of an image) to the GPU multiprocessor caches, then a kernel fragment program can be used to process the entire image, taking advantage of the massive computation power of the GPU. This processing manner especially allows for DLP exploitation.

2.5. Field programmable gate arrays

FPGAs are arrays of complex logic blocks with a network of programmable interconnects. They sprouted from the techniques of Programmable Logic Devices, such as Programmable Array Logic and Complex Programmable Logic Device. In general, FPGA architecture consists of logic blocks called configurable logic block, I/O blocks and routing fabrics (see Figure 2.4(a)). Its different logic combinations are realized according to look-up tables, whose outputs are fed to a D-type flip-flop in order to drive the other logic blocks or I/Os (see Figure 2.4(b)). In this way, a logic block available for both combinational logic and sequential logic functions is performed; one logic block is interconnected with each other or connected to I/O. In order to improve the performances and flexibility of current FPGAs, they are integrated by other different components, such as Block Random Access Memory, DSP48 (most commonly used to implement Digital Signal Processing operations with a maximum of 48-bits), or even processor blocks. They can be considered as customizable SoCs.

Due to their programmable nature, FPGAs can exploit different types of parallelism inherent in an image processing algorithm. With their high computational and memory bandwidth capabilities, FPGAs have already been used to solve many practical real-world, real-time image processing problems, from a preprocessing component to the entire processing chain.

In many cases, FPGAs have the potential to meet or exceed the performance of a single DSP or multiple DSPs. FPGAs can be considered as reconfigurable components combining the flexibility of software programmability with the speed of an ASIC within a shorter design or time-to-market.

In general, FPGA designs use some software programming languages known as Hardware Description Languages (HDL) to customize circuits. They provide precise execution times that help to meet hard real-time deadlines. Concretely, to define the behavior of FPGAs, the users first have to provide a HDL specification or a schematic design of the desired algorithm. Next, a technology-mapped netlist is generated according to an electronic design automation tool. Third, the netlist is fitted to the FPGA architecture via a place-and-route process. Finally, the user validates the map, place and route results via timing analysis, simulation and other verification methodologies. Once all of these processes are completed, the binary file generated is used to (re)configure the FPGA. We can note three levels of this process: Behavioral level described by HDL, synthesizable Register Transfer Level (RTL) and Gate level.

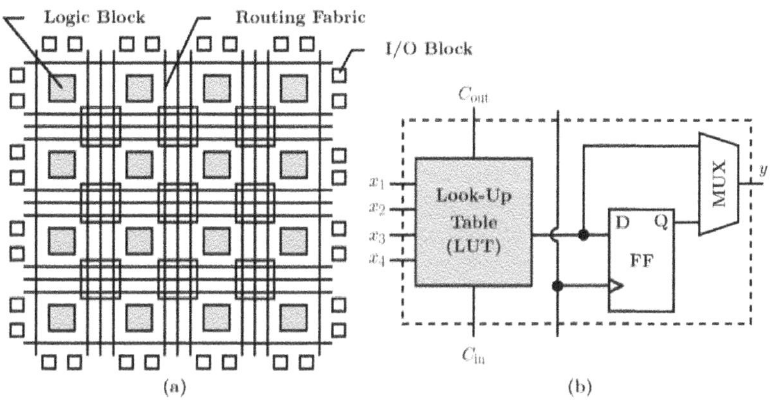

Figure 2.4. *Typical architecture of FPGA*

2.6. SW/HW codesign of real-time image processing

With the spreads of image processing techniques in different application fields and the emergences of variant computation platforms, the image processing development is today much more complicated than before, and transplanting such a design from its research environment to hardware environment becomes increasingly difficult and time consuming. In this context, the discipline "SW/HW codesign" appeared, which coordinates the work of the software and hardware engineers around the same objective but still keeps their efforts independent.

"Embedded computing is unique because it is a hardware-software co-design problem, the hardware and software must be designed together to make sure that the implementation not only functions properly but also meets performance, cost, and reliability goals" [CHI 94].

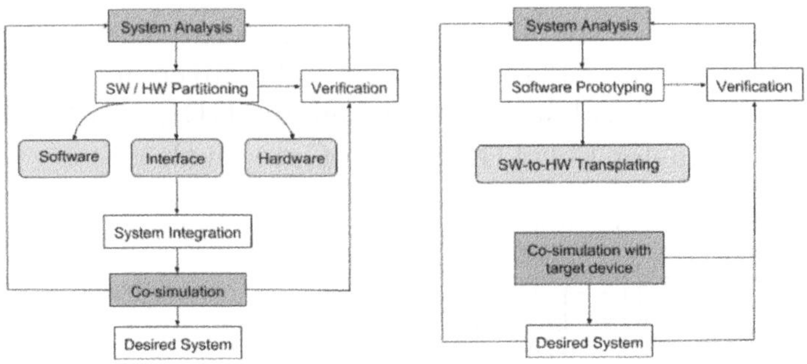

Figure 2.5. *Typical SW/HW codesign framework depending task partition modes: horizontal partition flow (left) and vertical partition flow (right)*

For a decade, a series of SW/HW codesign frameworks that coordinate the development of complex image processing applications have been in development for the purpose of high design productivity. We can distinguish two categories of development environments: those convenient for hardware implementation and those convenient for rapid algorithm prototyping.

The first category is those who horizontally partition the entire task (see Figure 2.5). First of all, the desired system is analyzed and the entire job is partitioned into two independent parts: software and hardware, which are dealt with by software and hardware engineers respectively. Next, software engineers prototype and functionally verify the software models, while hardware engineers construct the hardware platform. It should be noted that within this framework, the hardware part cannot only perform control flows, but also run some logic operations if necessary. This provides further potential opportunities to enhance the parallelism of designs. Finally, the two parts are synthesized together via a specific interface protocol.

The SW/HW codesign frameworks of the second category are those who vertically partition the task. This kind of design framework is based on the mature stand-alone device or development board, and what the engineers have to do is just to transplant the functionally verified algorithms from research environment to the target platforms. Depending on the design constraints, this translation can be made both manually and automatically. Finally, the hardware implementation of the design is evaluated through specific simulators for different devices.

2.7. Image processing development environment description

When computers had just been invented in the 1940s, either their architectures or the programs were quite simple. But with the development of computer sciences, the performance of processors has been increasingly improved, and the programs have also become more and more complicated. This resulted in the development efforts of programs becoming much more important than their quality. Thus, a series of integrated and visible programming environments were developed in order to increase programming efficiency, such as C, Pascal and FORTRAN.

Programming languages today can be divided into two types: assembly language and high-level programming languages. Assembly language evolved from machine language. It directly manipulates the hardware according to the abbreviated English identifiers. Its executable files generated from the source code are usually smaller and more efficient than the other languages.

Although assembly languages have a series of advantages in terms of file size and efficiency, high-level programming languages are the real favorite languages of programmers. High-level programming languages cannot only synthesize multiple concerned machine instructions into a single one, but also reduce the operations unnecessary for task specifications, such as stack and register manipulations. This approach effectively reduces the effort required for programming and the requirements of professional knowledge for the users.

Table 2.1 illustrates the compatibility evaluation of the development environments for different devices. In order to make an objective comparison, we classify the environment device compatibilities into four levels according to the following rules:

1) full compatibility (***): the programming languages are specially designed for the target devices, or their features are non-conditionally supported by the devices;

2) part compatibility (**): the programming languages are made available to the target devices through specific code translation tools, and insignificantly constrained by variances of hardware architectures;

3) low compatibility (*): the programming languages are seriously constrained by the hardware architectures, or their most important advantages cannot be profited by users due to code translation or hardware constraints;

4) non-compatibility (non-star): the programming languages are not available to the target devices.

	DSP	FPGA	GPP	GPU
Standard C/C++	***	**	***	
CUDA			***	***
OpenCL			***	***
Matlab		*	***	**
VHDL/Verilog		***		
LabVIEW/System Generator		**		

Table 2.1. *Compatibility evaluation of fundamental development environments for different hardware devices*

C/C++ is a general-purpose and imperative programming language. It provides constructs that efficiently map typical machine instructions and can be perfectly executed on DSPs and GPPs. Due to the fact that architectures of FPGAs are quite different from Von Neumann or Harvard architectures, C/C++-based FPGA designs have to be realized through a code translation tool. For GPUs, because standard C/C++ does not has the ability to specify kernel functions of GPUs, it cannot be used to configure GPUs.

CUDA and OpenCL are two parallel computing platforms and Application Programming Interfaces specially designed for GPUs or CPU-GPU heterogeneous platforms. Since parts of C language are incorporated, these two languages are also considered to be C-like languages, and can be easily mastered by C programmers.

Matlab is powerful commercial mathematical software developed by MathWorks. It provides users with an advanced computer programming language and interactive environment for algorithm development, data visualization and numerical analysis, etc. Although its main motivation is numerical operating, it can be as well used in variant fields such as system design, image processing, signal processing and communication modeling according to a series of different Toolboxes. Matlab language was originally designed for GPPs. In order to add to its flexibility, MathWorks equally provides multiple code translation tools to help users to automatically transplant their designs into other devices except for GPPs. For example, HDL Coder is used to generate HDL in RTL for FPGAs, while GPUArrays are used for GPU acceleration. However, due to the constraints of code translation strategies, arrays and vectors, which are the most frequently used variables in Matlab, are not supported by HDL Coder. Therefore, we consider the compatibility of Matlab with FPGA weak (single-star). Meanwhile, comparing with CUDA, the GPU map realized by Matlab is less efficient, so its compatibility is double-star.

VHSIC Hardware Description Language (VHDL) [IEE 11] and Verilog [IEE 06] are two hardware description languages specified for FPGA. They allow users to specify the hardware behavior in RTL. However, they are not convenient for algorithm development. Within VHDL or Verilog, users have to be capable of both software and hardware developments, and cannot concentrate their attentions on the algorithm aspect, which may bring more potential optimizations to the designs in terms of efficiency and accuracy.

Laboratory Virtual Instrumentation Engineering Workbench (LabVIEW) [INS 07] and System Generator [XIL 12a] are two visual programming languages for digital image processing. The motivation of these two languages is to improve the development productivity of FPGA designs by providing a higher abstraction environment than HDL. Generally, image processing system developments follow two main steps: schematic design and algorithm description. Within the visual programming languages, the users can built the desired systems using the components provided by the tool kits instead of describing the algorithm using textual programming languages. This object-oriented and procedure-oriented development method enables the users to finish their designs with few codes, thus effectively accelerating the development efficiency.

2.8. Comparison and discussion

In the previous sections, we review the currently available hardware platforms and development methodologies for real-time image processing designs. They include GPPs optimized for throughput over single-thread performance, DSPs optimized for compute-intensive digital signal processing and GPUs optimized for complex image processing to specific customizable chips, which are possibly implemented with reconfigurable hardware such as FPGAs.

These devices can often achieve better performance than conventional ones on certain workloads. However, applications usually exhibit vastly different performance characteristics depending on the target platform. This is an inherent problem attributable to architectural design, middleware support and programming style of the target platform. For the best application-to-platform mapping, factors such as programmability, performance, programming cost and sources of overhead in the design flows must all be taken into consideration.

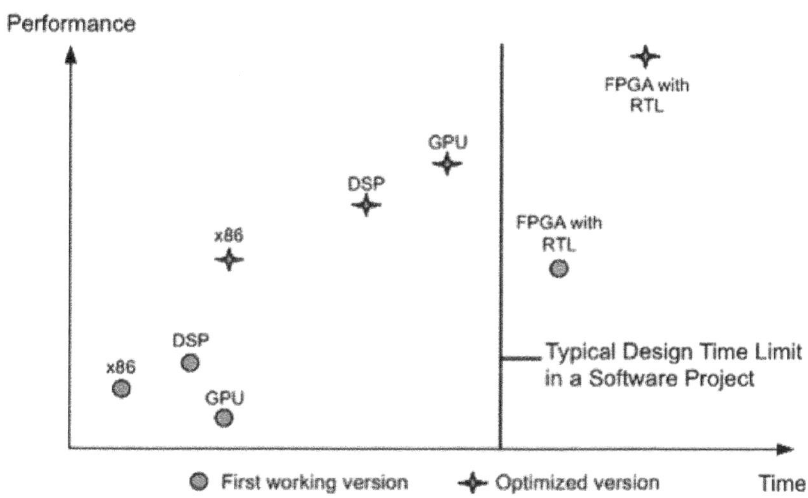

Figure 2.6. *Design times versus application performance with Register Transfer Level (RTL) design entry (see [XIL 13a])*

Recently, more and more image applications have been FPGAs as an implementation platform due to its flexibility, high memory bandwidth and real-time capacities. FPGA was originally considered a power consumption component in comparison with DSPs. Lately, low-power FPGAs are also becoming more available. In this context, a series of comparative studies indicate that FPGAs can often achieve better comprehensive properties than the others platforms in most cases. For example, in the work of Zou *et al.* [ZOU 12], the efficiency of the FPGA implementation of Smith-Waterman Algorithm is 3.4X to GPU and over 40X to CPU, while in the case of Kestur *et al.* [KES 10], it is indicated that FPGA has similar performance at higher energy efficiency when compared to the CPU and GPU platforms. In additional, the Advanced Digital Sciences Center of the University of Illinois at Urbana-Champaign reported that FPGAs can achieve a speedup up to 2–2.5× and save 84–92% of the energy consumption related GPUs [DEM 11].

However, FPGA is a reconfigurable device that has quite different hardware architecture than the others. It needs to be configured in the RTL before each use rather than run programs stored in the memories. This results in a fairly high effort cost for its application developments. According to the report of Xilinx [XIL 13a], despite higher performance, implementing an algorithm onto FPGAs usually requires a longer period than the other platforms. Figure 2.6 compares the time consumption of different platforms to develop the same software application. Both the initial and optimized versions of an application provide significant performance when compared against the same stages for both standard and specialized processors. RTL coding and an FPGA-optimized application result in the highest performance implementation. However, the development time required to arrive at this implementation is beyond the scope of a typical software development effort. The published data statistics in [ADS 16] corroborate this point: a manual FPGA design may consume 6–18 months and even years for a full custom hardware, while the GPU (CUDA) based designs only take 1–2 weeks. Che *et al.* [CHE 08] evaluate the performances of FPGAs and GPUs using some real-life applications; the final report concluded that:

– while computation involves a lot of detailed low-level hardware control operations, they cannot be effectively implemented using a high-level language;

– in order for the implementation to be able to take advantage of data streaming and pipelining, a certain degree of complexity is required;

– some applications require a lot of complexity in the logic and data flow design.

Despite higher performance, these comments demonstrate that RTL cannot provide an ideal software-convenient environment. Its code specifications force the engineers to pay more attention to low-level hardware description instead of software development. One of the consequences is that FPGAs were traditionally used only for mass-produced products or applications requiring a performance profile that could not be achieved by any other means.

One of the key factors that encourages the wide diffusion of FPGA devices is the improvement of man–machine interfaces (development environments), where the great challenge is to allow the easy use of complex electronic systems by non-hardware experts. Many research laboratories and industrial manufacturers are focusing their efforts to that effect.

3

Rapid Prototyping of Parallel Reconfigurable Instruction Set Processor for Efficient Real-Time Image Processing

This chapter describes research results on enabling the reconfigurable instruction set processor (RISP) model for real-time image processing applications by exploiting FPGA technology. The aim is to keep the flexibility of processors in order to shorten the development cycle and at the same time profit from the powerful FPGA resources to increase real-time performance. Using advanced compiler technology, we designed modular RISP processor VHDL models with a variable instruction set and a customizable architecture which makes it possible to exploit the intrinsic parallelism of a target application and implement it in an optimal manner onto an FPGA. We illustrated the proposition by using the 2D spatial Cosine Transform algorithm with a variable block size on a Virtex-6-based board. Hardware implementations were realized in several ways by using either chip area or speed optimizations. Experimental results show that this approach applies some criteria to codesign tools: flexibility, modularity, performance and reusability.

3.1. Context and problematic

Adaptability of computing systems to support various multimedia formats and equipment with different computational requirements is highly desirable in the ubiquitous and seamless computing era. Currently, more and more image applications are choosing FPGAs as an implementation platform due to their flexibility and low power consumption. Aside from their massive parallelism capabilities, multimillion gate counts and many built-in circuit features, the most interesting characteristic of an FPGA is probably the ability to quickly create a rapid and fully functional prototype that can emulate and verify solutions or even be embedded into the final system.

In the classical FPGA-based prototyping methodology, the application source code is first translated into a hardware description language such as VHDL or Verilog. Then, the design is synthesized for an FPGA-based environment where it can be tested. Most hardware description languages are naturally inherently concurrent and non-hardware developers do not consider most low-level hardware languages to be trivial. One of the key factors that encourages the wide diffusion of electronic devices is the improvement of tool suite interface conviviality, where the great challenge is to allow for the use of complex electronic systems by software developers. Many research laboratories and industrial manufacturers are focusing their efforts on this aspect [AGU 05, BRO 07]. Two major trends emerge.

The first trend, based on the fact that software developers know much more about traditional software high-level programming languages such as C and C++ than about hardware description languages, consists of extending these languages with variations capable of describing hardware elements [PAN 01, LOO 02, TIE 05]. These recent hardware description languages are, in effect, parallel synchronous programming languages where the notion of time is fundamental to its specification. In these languages, all events occur relative to a global clock that runs continuously. Information is encoded on a behavioural level in a similar manner to most high-level languages.

The second trend consists of using programmable devices, a processor core being included in an FPGA for example. Some parts of an application can be programmed with high level languages such as C or C++. The rest of the FPGA is used to manage communication peripherals or integrate low-level processing blocks. The soft-core processor, introduced in the last decade, is a microprocessor fully described in software, usually in VHDL, and capable of being synthesized in hardware, such as an FPGA. For example, the two major FPGA manufacturers provide commercial soft-core processors. Xilinx offers its MicroBlaze processor [XIL 09a], while Altera has Nios and Nios II processors [ALT 16]. All these programmable devices including hard or soft-cores can be considered single chip implementations of an SW/HW hybrid system [AGU 05].

Our approach combines these two trends. For the rapid prototyping of image processing applications on the FPGA, algorithms are programmed in C as if they were to be executed on a classical CPU. The source code is analyzed automatically and an RISP VHDL model is generated and implemented with FPGA technology. A specific tool detects and extracts intrinsic instruction level parallelism (ILP) in order to improve real-time performance.

This development framework performs design exploration of RISP, which incorporates custom functional units within the processor architecture. This rapid prototyping approach provides an efficient mechanism to meet the growing performance and time-to-market demands. In particular, we can effectively generate reconfigurable functional units (RFUs) which only contain the minimum number of instruction sets or modules, just necessary for each target application. We call this a minimum mandatory modules (M^3) methodology. Thus, several RISPs can easily be made with a single FPGA in order to build a high-level parallel machine.

3.2. Related works

A RISP consists of a microprocessor core that is tightly coupled with one or several RFUs [BAR 02]. It possesses a higher degree of customization to manage the growing complexity of the applications. At the same time, its high degree of flexibility allows it to meet the shrinking time-to-market window.

With its reconfigurable nature, FPGA greatly facilitates RISP development as the design flexibility of RISPs in the presence of these reconfigurable logics leads to off-the-shelf products that can be customized for each application. This is increasingly preferred by designers who develop products for uncertain markets and shorter product life cycles [LU 08, JUN 11]. For example, rapid reconfigurability is especially attractive for applications in ubiquitous computing with evolving standards, which require frequent functionality updates [GLE 04].

The major challenges to increase the proliferation of RISPs lie in the development of supporting compilation and computer-aided design tools which enable rapid design exploration and efficient mapping of applications on such platforms [BIS 06, HAS 11, CHE 07, SHI 08]. In [LAM 09], Lam and Srikanthan present a framework that enables rapid design of area-efficient custom instructions for reconfigurable embedded processing, which incorporates a four-wide very long instruction word (VLIW) RISP structure. In particular, it provides rapid identification of a reduced set of profitable custom instructions and their area costs on commercial architectures without needing a time consuming hardware synthesis process. Obtained experimental results for the generic, automotive, industrial, image, network, security and telecommunication application sets show that the number of candidates for custom instruction selection can be significantly reduced by 30–70% with marginal degradation in the resulting performance gain.

In a similar manner, Seto and Fujita [SET 10] presented a novel framework generating efficient custom instructions for common configurable processors with limited numbers of I/O ports in the register files and fixed-length instruction formats, such as RISCs. Their work generates a sequence of multiple custom instructions by applying high-level synthesis techniques such as scheduling and binding to the subgraphs. In comparison with some previous works, this approach provides the following advantages simultaneously: (1) generation of effective custom instruction from general multi-inputs multi-outputs subgraphs without changing the pipeline and the instruction format of the configurable processors and (2) generation of a single, area-efficient custom functional unit for a set of custom instructions in which resources are shared among different custom instructions. An image processing algorithm set (fft, dct, djpeg) has been implemented on the Altera Stratix II FPGA device using the proposed framework and this greatly improves performances in the target topic.

We can also cite the works realized by Kariri *et al.* [KAR 08] concerning a design flow for architecture exploration and implementation of reconfigurable application-specific instruction set processors (rASIPs). In order to deliver excellent balance between performance and flexibility, their design flow is centered on all three major components of a rASIP: the programmable core, the reconfigurable fabric and the interfaces between these two through automatic generation of the software tools (compiler tool chain and instruction set simulator) and the RTL hardware model.

Several case studies in image/video processing have been tested using the proposed framework.

In previous works, we have proposed a framework in order to implement a VLIW DSP TMS320C62xx of Texas Instrument (TI) onto the FPGA for real-time signal and image processing [BRO 07]. With this approach, we easily exploit the parallelism of target applications using the VLIW processor compiler (Code Composer Studio of TI) and quickly implement corresponding functional units onto the FPGA using a DSP VHDL model generator. The most innovative aspect of our work lies in the concept of a modular VLIW processor model (M^3 methodology).

In this chapter, we want to generalize the M^3 methodology to a general RISP processor with generic kernel. The instruction length is modular according to the target application such as the number and type of functional units. We use the LLVM compiler infrastructure [LLV 17] to convert the C/C++ program source into intermediate representation (IR) in order to exploit intrinsic ILP. First, we introduce a specific tool to analyze and expose all operations that can be executed concurrently in order to shorten the application execution cycle number to a minimum. Secondly, a RISP VHDL processor is automatically generated, with minimum mandatory modules for target application. Finally, all VHDL files are synthesized and implemented onto FPGA. We can thus constitute a parallel RISP set that is optimal for each target application.

3.3. Design exploration framework

The proposed framework in Figure 3.1 consists of three stages: (1) Sequential pattern file generation, (2) Parallel pattern file generation, and (3) Hardware generation. Its input corresponds to an application programming in classical and standard C language and the output to FPGA implementation.

In the sequential pattern file generation stage, the LLVM compiler infrastructure [LLV 17] is used to convert C program into IR. The LLVM project is a collection of modular and reusable compiler technologies in order to provide a modern, SSA-based compilation strategy capable of supporting both static and dynamic compilation of arbitrary programming languages [BER 10, HUA 10, GEO 10]. We use LLVM as a code generator in order to establish a well-specified register-based assembly representation. Then, we enumerate the custom instruction pattern instances from the application's IR code. For each run instruction, we note used registers for operands and operation type. At the end of this stage, a sequential execution graph is built with the totality of necessary registers and operation types, and these pattern instances are stored in the sequential pattern file.

The parallel pattern file generation stage is divided into two steps: pattern analysis and pattern matching. First, from the sequential pattern file, we browse the sequential execution graph in order to (1) examine data dependency for each the cycle, (2) extract ILP, under some predefined constraints (simultaneous memory access, maximum number of units, etc.) and (3) enumerate custom instructions that can be executed in parallel. Then, these reorganized pattern instances are stored in the parallel pattern file.

In the parallel pattern matching step, we perform optimizations to reduce the used register number and choose the best compromise between the maximal number of operations that can be realized in parallel and the minimal necessary number of parallel cycles for the target application. This allows us to establish the optimal parallel pattern file.

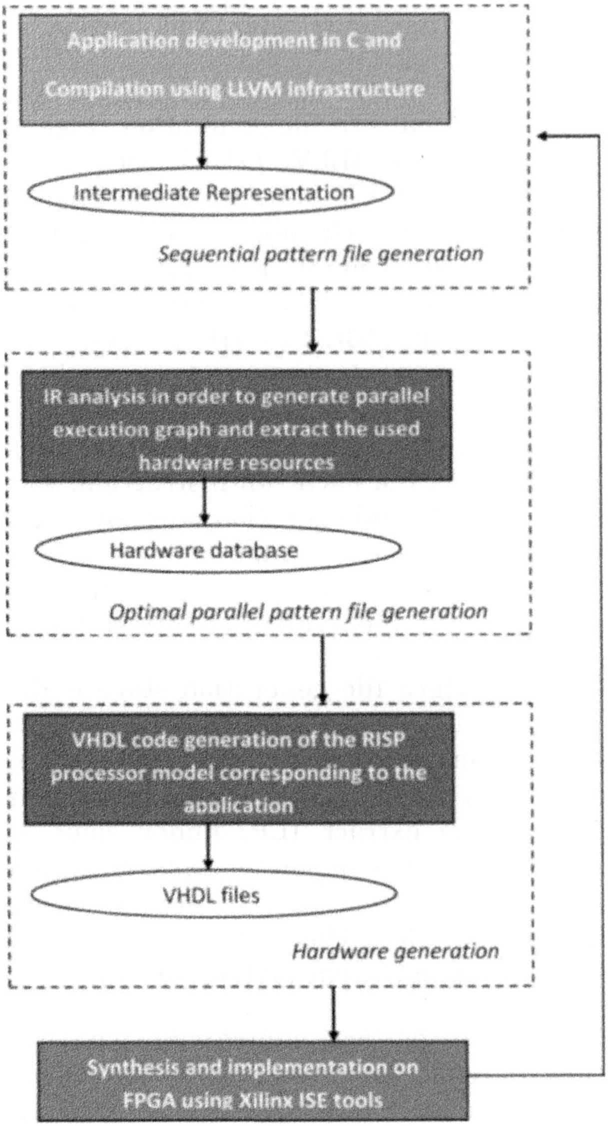

Figure 3.1. *Overview of the rapid prototyping framework*

The goal of the third stage is to generate the RISP processor structure, coupled with one or several RFUs. In this hardware generation stage VHDL files are automatically generated and synthesized in agreement with optimal parallel pattern files. It begins with the enumerating of RFUs and registers. Then, we design the RISP with three VHDL files: *RFUs.vhd*, *Registers.vhd* and *ROM.vhd*. Figure 3.2 shows the generic kernel of RISP in which each RFU possesses four inputs with three operands of 32-bits and an operation type selection, and an output corresponding to operation result. These inputs/outputs are connected with used registers that are described in *Registers.vhd*. The *ROM.vhd* file can be considered as program code that performs, for each parallel cycle, activation of RFUs used for this cycle with correct selection of registers for operands, result, and associated operation type.

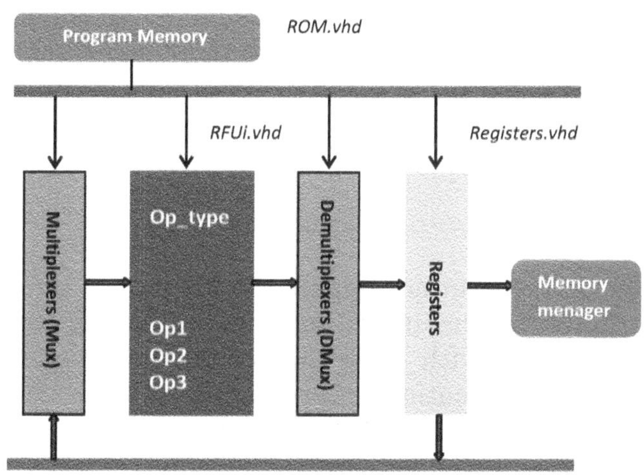

Figure 3.2. *Illustration of the generic kernel structure of RISP*

Note that the instructions available for our RISP processor model consist of the basic instruction set used by all classical RISC processors as arithmetic and logic operations, control flow operation and two memory operations (read/write). The instruction set of each, RISP based on a target application, is first generated by the LLVM compiler, and then reorganized and optimized by our tool suite. These instructions already correspond to a subset of available LLVM instructions [LLV 17].

3.4. Case study: RISP conception and synthesis for spatial transforms

In this section, we illustrate the RISP rapid prototyping framework using a spatial transform algorithm with a variable block size. Discrete cosine transforms (DCT) possess a strong decorrelation and energy compaction property. So, many image and video compression standards such as JPEG, MPEG, H.264/AVC and high-efficiency video coding (HEVC) use DCT to transform the image data from the spatial domain to the frequency domain. In the following sections, we first describe two digital implementations of the 2D-DCT algorithm, and then show their corresponding RISP conception and optimization before presenting the results of simulation and synthesis.

3.4.1. *Digital DCT algorithm implementations*

The 2D-DCT operation reduces the amount of information used to represent the image or video, allowing for higher compression rates with little impact on the image quality, because the information represented in the frequency domain can be processed to discard the frequencies that are less important to the human visual system's perception. The DCT transform can be achieved by matrix operation or butterfly structure [LEE 97, CHE 77, LOE 89].

The 2D-DCT and IDCT of an $N \times N$ block are defined as:

$$F(u,v) = \frac{2}{N} C(u)C(v) \sum_{m=0}^{N-1} \sum_{n=0}^{N-1} f(m,n) \cos\frac{(2m+1)u\pi}{2N} \cos\frac{(2n+1)v\pi}{2N} \qquad [3.1]$$

$$u,v = 0,1,\ldots,N-1$$

$$f(m,n) = \frac{2}{N} C(m)C(n) \sum_{u=0}^{N-1} \sum_{v=0}^{N-1} F(u,v) \cos\frac{(2m+1)u\pi}{2N} \cos\frac{(2n+1)v\pi}{2N} \quad [3.2]$$

$$m,n = 0,1,\ldots,N-1$$

where $F(u, v)$ is the DCT coefficient and $f(m, n)$ is input data, $C(0) = 1/\sqrt{2}$, $C(j)=1$ for $j \neq 0$. From the definition, we can note that the basic structure of DCT and the one of IDCT are the same. The integer transforms can be represented as [CHA 89]:

$$F_N = (I_N \times f_N \times I_N^T) \otimes E_N \quad [3.3]$$

where I_N is the $N \times N$ integer transform matrix, E_N is the scaling matrix, "T" is the matrix transport operation and $\otimes$ denotes term-by-term multiplication, which is usually approximated by integer multiplication and shift.

The 2D-DCT/IDCT can be calculated by repeatedly applying the 1D transforms: first 1D-DCT/IDCT of each row, followed by 1D-DCT/IDCT of each column (or vice versa). Here, we use $N = 8$ to show the matrix operation and butterfly implementation of 1D-DCT. When $N = 8$, the integer transform matrix can be expressed as:

$$I_8 = \begin{bmatrix} a & a & a & a & a & a & a & a \\ b & d & e & g & -g & -e & -d & -b \\ c & k & -k & -c & -c & -k & k & c \\ d & -g & -b & -e & e & b & g & -d \\ a & -a & -a & a & a & -a & -a & a \\ e & -b & g & d & -d & -g & b & -e \\ k & -c & c & -k & -k & c & -c & k \\ g & -e & d & -b & b & -d & e & -g \end{bmatrix} \quad [3.4]$$

To realize the DCT transform by matrix operation, we compute:

$$
\begin{aligned}
F_0 &= a\big((f_0+f_7)+(f_3+f_4)\big)+a\big((f_1+f_6)+(f_2+f_5)\big)\\
F_4 &= a\big((f_0+f_7)+(f_3+f_4)\big)-a\big((f_1+f_6)+(f_2+f_5)\big)\\
F_2 &= c\big((f_0+f_7)-(f_3+f_4)\big)+k\big((f_1+f_6)-(f_2+f_5)\big)\\
F_6 &= k\big((f_0+f_7)-(f_3+f_4)\big)-c\big((f_1+f_6)-(f_2+f_5)\big)\\
F_1 &= b(f_0-f_7)+d(f_1-f_6)+e(f_2-f_5)+g(f_3-f_4)\\
F_5 &= e(f_0-f_7)-b(f_1-f_6)+g(f_2-f_5)+d(f_3-f_4)\\
F_3 &= d(f_0-f_7)-g(f_1-f_6)-b(f_2-f_5)-e(f_3-f_4)\\
F_7 &= g(f_0-f_7)-e(f_1-f_6)+d(f_2-f_5)-b(f_3-f_4)
\end{aligned} \qquad [3.5]
$$

Equation [3.4] shows that there is a symmetric property between the left side and the right side of the transform matrix. Based on this property, the DCT can be also implemented by a fast algorithm called butterfly. In the calculation equation [3.5], there are many overlaps in the eight equations. These overlaps can be used to reduce the computation complexity. For example, after combination, the f_0+f_7 is used in four equations, so this term needs to be calculated only once. Based on the combination, the multiplication times are also reduced in each equation. Figure 3.3 displays the final partial butterfly flow graph corresponding to 1D-DCT of block size $N = 8$. The matrix operation and butterfly structure of the DCT and IDCT transform are all implemented in the HM5.2 [SAM 10] test model.

3.4.2. *Rapid prototyping of DCT RISP conception*

We use the 1D-DCT partial butterfly operation with $N = 4$ as an example to illustrate the RISP conception steps. First, the C source is provided to the LLVM compiler in order to translate it into assembly IR. In the ILP extraction and optimization step, we analyze data and operation dependences to transform the original sequential assembly code in parallel basic operations to generate the final parallel execution graph.

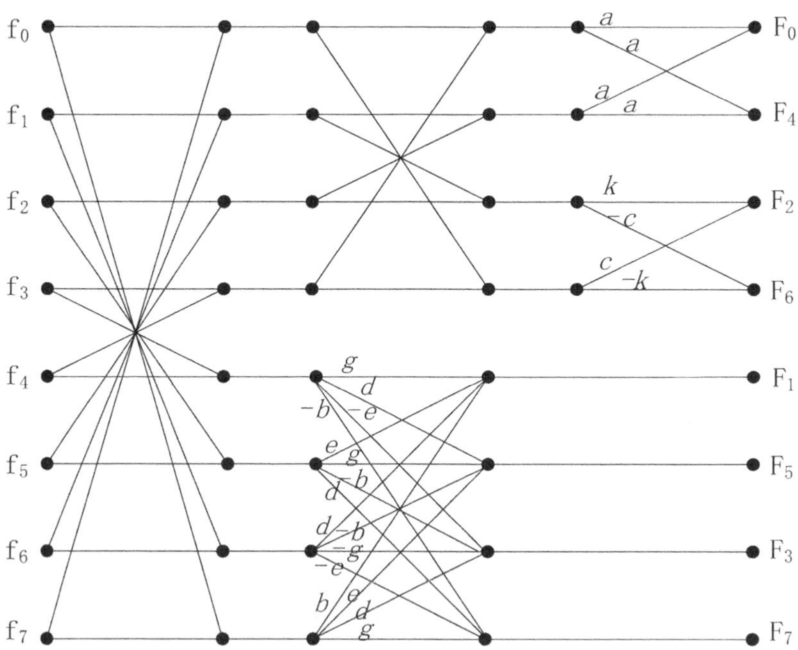

Figure 3.3. *Eight-point DCT partial butterfly flow graph*

Figure 3.4 displays the parallel execution graph generated by our software tool for this 1D-DCT processing. We can see that the main loop was performed using eight cycles after parallelization. Here, *Ri* (i = 0–6) corresponds to a register and *load* (*store*) to a data loading (storing) from (to) memory. The *shl (shr)* represents a data shift left (right). Red arrows indicate constants. For example, at step (cycle) 4, seven operations were simultaneously realized: three additions, three subtractions and one shift left. This parallelization allows us to accelerate calculations with a ratio of 3.71 which corresponds to number of sequential cycles per number of parallel cycles.

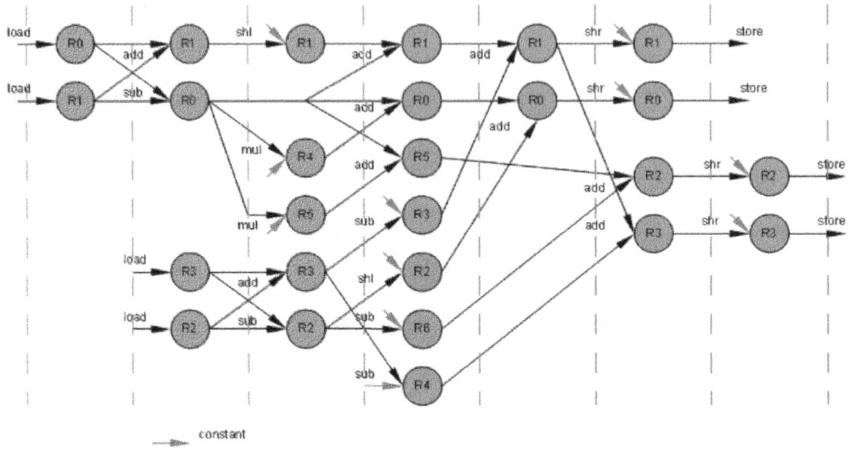

Figure 3.4. *Parallel execution graph generated for the 1D-DCT partial butterfly operation: each operation is realized using two operands from two registers and memorizes the result in a register. For a color version of the figure, see www.iste.co.uk/li/image.zip*

In the same manner, we design several generic kernels of RISP, and all these reprogrammable processors are capable of processing spatial transforms with a variable block size. Usually, the 2D-DCT is realized with block size $N = 4$ or 8. However, compression efficiency can be significantly improved by utilizing macroblock structures with sizes larger than 8 × 8 pixels, especially at high resolutions [MA 07, UGU 10]: there are more likely to be large homogenous areas that can be efficiently represented by larger block sizes. For example, the last HEVC standard is able to perform DCT transform on sizes from 4 × 4 to 32 × 32.

For design space exploration (DSE) purposes, we conceived these RISPs in four ways:

– matrix operation: DCT and IDCT transforms have been directly implemented by multiplication calculations using an integer transform matrix;

– matrix operation with loop iteration reduction option "4": we used a mechanism of loop unrolling present in the LLVM compiler optimization option. In fact, loop overhead can be reduced by reducing the number of iterations and replicating the body of the loop; this technique is called loop unrolling or loop unwinding (see more details in section 5.2.4). In this way, the reduction factor of loop iteration is equal to 4;

– matrix operation with loop iteration reduction option "8": the reduction factor of loop iteration is equal to 8;

– partial butterfly operation: DCT and IDCT are implemented using the fast partial butterfly algorithm described in the previous section.

In the following presentation, these four RISP processors are called, respectively, MX, R4, R8 and PB. Table 3.1 shows obtained acceleration values for these 2D-DCT RISP processors after parallelism extraction. We can observe that acceleration varies from 1.84 to 7.24 according to block size and implementation path.

	MX	R4	R8	PB
$N = 4$	1.84	3.51	4.14	2.05
$N = 8$	1.92	4.37	5.54	2.09
$N = 16$	1.96	4.93	6.59	2.11
$N = 32$	1.98	5.25	7.24	2.12

Table 3.1. *IR assembly and parallelism extraction results: acceleration corresponds to number of sequential cycle per number of parallel cycle for one block of 2D-DCT processing*

3.4.3. *RISP simulation and synthesis for 2D-DCT*

After the parallelism extraction and optimization step, the target RISP architecture (RFUs and registers) and its instruction set are configured. Then, we automatically generate a VLIW RISP processor VHDL model for four ways: MX, R4, R8 and PB. Note that our RISP processor is totally customizable and optimal for target realization: number of functional units, operation types and registers used are the minimum necessary for given operation.

We use the Xilinx ISE [XIL 09b] simulator for FPGA-based implementation and the target hardware architecture is the Virtex 6-vlx 240 [XIL 09c]. It contains 768 DSP48 slices (with 25 × 18 multipliers and a 48-bit adder/subtracter/ accumulator), which support massively parallel digital signal processing algorithms, 416 intern block RAMs and 37,680 configurable logic blocks (CLBs). Slices of the CLBs can be used to provide logic, arithmetic and ROM functions; part of them can also be used for distributing RAM or 32-bit data registers.

Figure 3.5 displays the necessary cycle number of execution for each RISP, which corresponds to one block spatial transform processing. Four block sizes were simulated: N= 4, 8, 16 and 32. We also give used hardware resources of the target FPGA in Table 3.2.

We can observe that the PB RISP is more speed optimized with a lower cycle number for 16 × 16 and 32 × 32 block sizes. The MX RISP is more chip area optimized, because of the minimum used hardware resources. Note that the RISP R4 reduces the cycle number by more than half in comparison with the MX in all four size cases. It uses only one quarter of hardware resources compared to the PB RISP.

In order to make comparisons in terms of ratio area/speed, for four block sizes (N = 4, 8, 16 and 32), we performed normalization operations using a vector norm. Figure 3.6 shows the products of normalized cycle number and normalized slice number of each RISP. These curves confirm that the PB algorithm is only interesting for large block sizes

(N = 16 and 32). The R4 RISP constitutes the best compromise between processing speed and hardware resources used. In fact, the loop unrolling operation can be considered as task distribution processing in the spatial transform case. For all block sizes from 4 × 4 to 32 × 32, division of processing by four tasks allows us to reduce the cycle number of execution using task parallelism with better ILP extraction. On the other hand, the R8 RISP realized task parallelism by 8; this way, there are more RFUs compared to matrix operation needs, which results in a poorer performance.

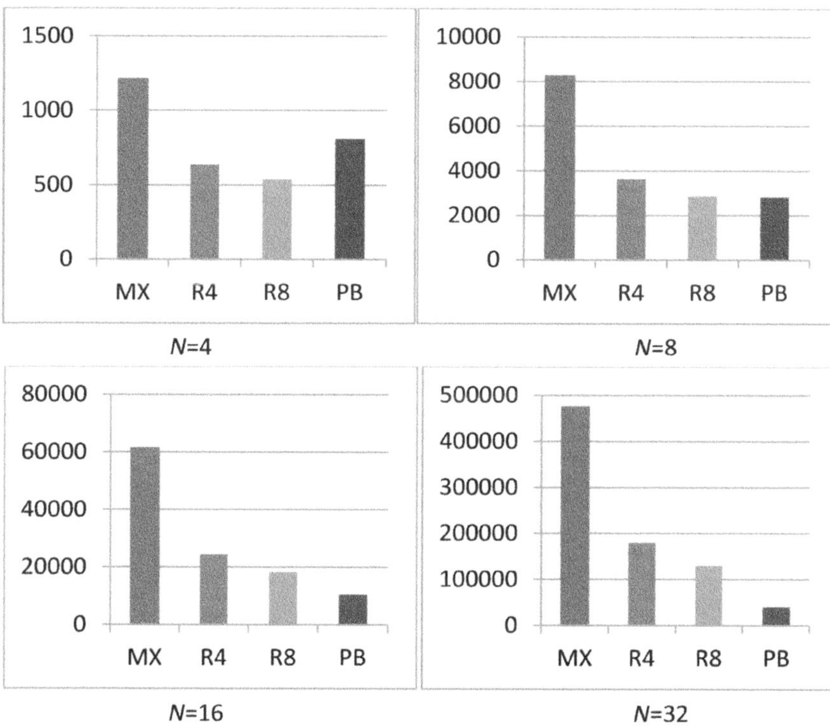

Figure 3.5. *Necessary cycle number of each RISP to perform spatial transform of one block. Four RISPs correspond to MX, R4, R8 and PB. Block size is equal to 4 × 4, 8 × 8, 16 × 16 and 32 × 32. For a color version of the figure, see www.iste.co.uk/li/image.zip*

	MX	R4	R8	PB
CLB Slices	839	1794	5838	7563
DSP48	1	2	2	7
Freq. (MHz)	184	186	171	170

Table 3.2. *Used hardware resources and operating frequency of each RISP processor*

There remain two remarks to specify concerning RISP enabling on the FPGA component:

– based on the fact that recent FPGAs all possess specific multipliers, or even DSP elements, we did not try to suppress multiplication operations in the spatial transform;

– in all RISP processors, we respected the hardware component constraint: the operating frequency of modern memories (e.g. 800 MHz for DDR3) can achieve at least two accesses per cycle of RISP.

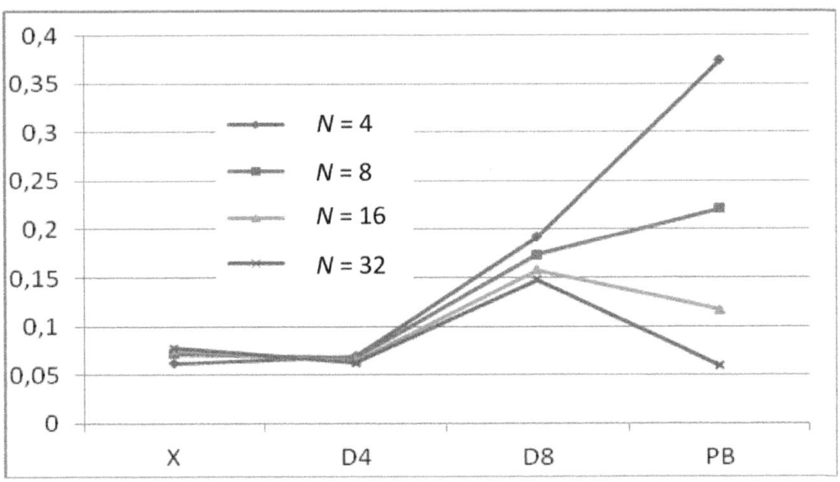

Figure 3.6. *Products of normalized slice numbers and normalized cycle numbers: a small product value means a better area/speed ratio. For a color version of the figure, see www.iste.co.uk/li/image.zip*

3.5. Hardware implementation of spatial transforms on an FPGA-based platform

We used the ML605 evaluation board [XIL 09c] as the hardware platform for embedded spatial transforms. It enables hardware and software developers to create or evaluate designs targeting the Virtex-6 XC6VLX FPGA. The ML605 provides board features common to many embedded processing systems, which include a DDR3 SODIMM memory, an 8-lane PCI express interface and a UART.

Figure 3.7. *Four classical test video sequences: Foreman (CIF), Race horse (WQVGA) , Tennis (HDTV) and Akiyo (QCIF). For a color version of the figure, see www.iste.co.uk/li/image.zip*

In order to test and validate our spatial transform RISP processors, we chose four classical and standard video sequences with different formats: Foreman, Race horse,

Tennis and Akiyo (see Figure 3.7). Using decoder codes of the HEVC test model HM5.2 [SAM 10], we performed a decoding process to discover the macroblock distribution of these video sequences. Table 3.3 gives percentages of four block sizes for each sequence in the HEVC decoding process. These results correspond to the average value of 100 frames. There is a small percentage of large blocks for the first three sequences, with QCIF, CIF and WQVGA formats. On the other hand, fewer than half of the blocks of the Tennis sequence (HDTV format) correspond to the 4 × 4 size.

Block size	4 × 4	8 × 8	16 × 16	32 × 32
Akiyo (144 × 176)	76%	21%	2.6%	0.4%
Foreman (288 × 352)	77%	20%	2.9%	0.3%
Race horse (242 × 416)	85%	13%	1.8%	0.2%
Tennis (1080 × 1920)	47%	31%	17%	5%

Table 3.3. *Macroblock distribution of four video sequences using the HEVC decoding process*

Figure 3.8 illustrates the block diagram of hardware implementations of spatial transforms onto the ML605 evaluation board. We give a brief description of each block:

– MicroBlaze [XIL 09a] is an embedded soft-core processor optimized for Xilinx FPGAs. It is designed to be flexible and provides control of a number of features: the 32-bit RISC Harvard style architecture with separated instruction and data buses running at full speed to execute programs and access data from both on-chip and external memory. According to the level of performance desired, the number of slices used varies from 1,860 to 2,254 for an operating frequency between 170 and 220 MHz on Virtex 6;

– memory controller is an intellectual property (IP) block provided by Xilinx in order to connect the MicroBlaze with the external memory. Here, we only use three ports (four available): two for the RISP processor and one for the MicroBlaze;

– the PLB bus constitutes an interface of MicroBlaze with external peripherals. This IP-block is a standard bus of systems-on-chip. The MicroBlaze drives the RISP processor via this bus: register initiation, execution start and reset at program execution end. We can also use debug mode;

– the USB2 controller establishes the interface of hardware platform with PC. For example, test data stored in a DDR3 memory are updated via this IP-block.

NOTE.– In electronic design, an (IP) block/core is a reusable unit of logic; cell or chip layout design that can be used as building blocks within ASIC chip designs or FPGA logic designs.

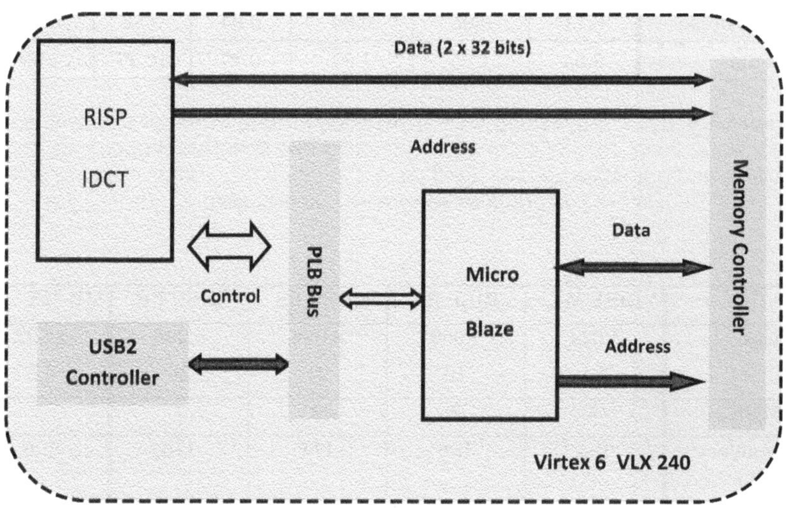

Figure 3.8. *Block diagram of hardware implementations of spatial transform on the ML605 evaluation board*

Using the rapid prototyping of RISP on the FPGA framework described in section 3.3 and simulation results obtained in section 3.4, we implemented four IDCT RISP onto the ML605 platform: MX, R4, R8 and PB. In order to make some comparisons, we also performed IDCT processing on the MicroBlaze in two manners: matrix operation and partial butterfly. This soft-core processor can only execute instructions sequentially. All hardware implementation results are displayed in Tables 3.4 and 3.5. The RISP can be considered as an accelerator; for example, in comparison with MB (MicroBlaze)–MX (matrix operation), we obtained an acceleration of 3×, 9× and 10× respectively using RISP-MX, RISP-R4 and RISP-R8. For the partial butterfly algorithm, only an acceleration factor of 2× is observed.

	RISP MX	RISP R4	RISP R8	RISP PB	MB MX	MB PB
Akiyo	**0.03**	**0.01**	**0.01**	**0.01**	0.1	**0.02**
Foreman	0.12	0.05	**0.04**	**0.04**	0.47	0.08
Race horse	0.11	0.05	**0.04**	**0.04**	0.43	0.08
Tennis	5.05	1.95	1.57	0.77	18.77	1.59

Table 3.4. *IDCT processing time in seconds for one frame of different video formats: these results were obtained using five soft-core processors (four RISPs and one MicroBlaze). Bold values indicate that the processing speed is equal or superior to 25 frames/s*

	RISP MX	RISP R4	RISP R8	RISP PB	MB MX - PB
Slices	3,074	4,030	8,074	9,798	2,592
%	8.1	10	21	26	6.8
DSP48	1	2	2	7	3
Frequency (MHz)	184	186	171	170	196

Table 3.5. *IDCT processing used hardware resources and operating frequency: these results were obtained by using five soft-core processors (four RISPs and one MicroBlaze)*

For small video resolutions, IDCT of the HEVC can be realized in real time. Note that our main goal consists of designing thus rapid prototyping of image/video processing on FPGA, thus reducing development time to adapt time-to-market constraints. We did not try to manually accelerate these implementations using specific hardware knowledge. In comparison with the MicroBlaze, RISP-R8 and RISP-PB use more than 20% of the available Slice resources. The RISP-R4 is always a better compromise between the processing speed and the used chip area.

3.6. Discussion and conclusion

In this chapter, we present the results of embedded HEVC spatial transforms using a novel framework that enables rapid design exploration for RISP, which incorporates custom functional units within the processor architecture. Our goal is not only to achieve high performance of hardware implementation in real time, but also to provide a new rapid prototyping tool suite to software/hardware developers. For example, with a FPGA or application, we can perform rapid DSE in order to simultaneously respect algorithmic and hardware constraints due to the adaptive capacity of our approach. For example, using area optimization, the RISP-MX of spatial transform can be used in mobile devices where power and chip area is the main concern. The speed optimized RISP-PB can be in high-definition encoders where the joint data flow of transform and predictor/residue generation is considered.

In comparison with related works, we concentrate our efforts on low-level parallelism extraction and minimal hardware model generation. In the image/video domain, processing full high-dimension images leads to millions of identical independent operations in the different parts of the image. In this context, ILP extraction is particularly important and efficient before DLP exploration. This design framework, for the first time, provides multiple advantages:

1) it possesses more programming flexibility: its input consists of classical and standard C or C++ source code. Because of the automatic generation of the RISP VHDL processor model, software engineers can realize rapid prototyping of signal and image processing on FPGA devices, completely ignoring hardware aspects;

2) it realizes performance-efficient hardware implementations: the concept of generic kernel of the RISP processor allows us to exploit intrinsic ILP to the maximum to reduce the necessary execution cycle number. At the same time, the generated RISP VHDL processor model just uses the minimum necessary hardware resources for each target application. This results in better performance in terms of area/speed ratio;

3) it is an open and adaptive framework that eventually allows several level parallelisms to be exploited (ILP, DLP and task pipeline, etc.), due to the economic use of available hardware resources. On the other hand, with a given FPGA or application, we can perform rapid DSE in order to simultaneously respect algorithmic and hardware constraints due to the adaptive capacity of the tool suite.

In conclusion, we also think that rapid design of RISPs is very appropriate for emerging and evolutionary standards and algorithms.

4

Exploration of High-level Synthesis Technique

High-level synthesis (HLS) is a promising technique to increase FPGA development productivity. It allows users to reap the benefits of hardware implementation directly from the software source code specified by C-like languages. This chapter first explores the synthesis process of Vivado_HLS, one of the leading HLS tools. Then, we implement a novel high-convergence-ration Kubelka–Munk genetic algorithm (HCR-KMGA) in the context of a multispectral skin lesion assessment in order to evaluate the technique.

4.1. Introduction of HLS technique

HLS is also called C synthesis or electronic system level synthesis. It allows the synthesis of the desired algorithm specified using the HLS-available C language into RTL and performs the RTL ports connection for the top behavior with specific timing protocol depending on the function arguments. The motivation of this technique is to enable hardware designers to efficiently build and verify their targeted hardware implementations by giving them better control over the optimization of their design architecture and allowing them to describe the design in a higher abstract level.

Despite a series of failures of the commercialization of HLS systems (since 1990s), the innovative and high-quality HLS solutions are always strongly required due to the wide spread of embedded processors and the increasingly fierce market competitions among the electronics manufacturers. Cong *et al.* [CON 11] summed up the motivations of HLS from the following five aspects:

– wide spread of embedded processors in system-on-chips (SoCs);

– huge requirement of higher level of abstraction from the huge silicon capacity;

– benefits of behavioral IP reuse for design productivity;

– verification drives the acceptance of high-level specification;

– trend toward extensive use of accelerators and heterogeneous SoCs.

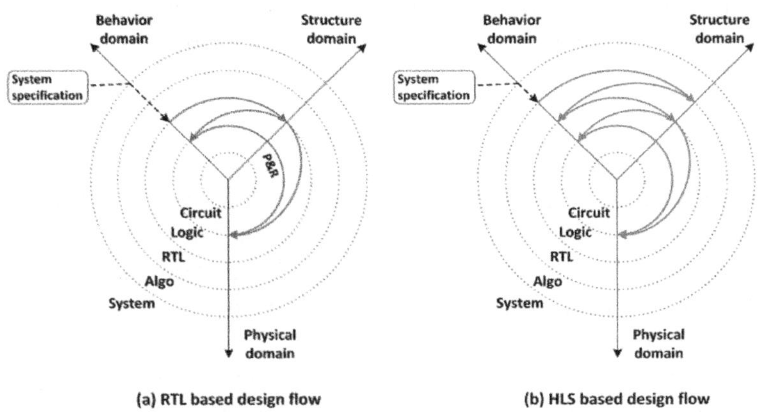

Figure 4.1. *Comparison of RTL- and HLS-based design flows by using Gasjki-Kuhn's Y-chart: full lines indicate the automated cycles, while dotted lines the manual cycles*

HLS is an effective method to reduce the R&D time by automating the C-to-RTL synthesis process. Figure 4.1 compares the conventional RTL with the HLS-based design flows by using Gasjki-Kuhn's Y-chart [MEE 12]. We can see that the philosophy of this method is to provide an error-free path from abstract specifications to RTL. Its benefits include:

– reducing the design and verification efforts and helping for investing R&D resources where it really matters. When working at a high level of abstraction, a lot less detail is needed for the description, i.e. hierarchy, processes, clocks or technology [LIA 12]. This makes the description much easier to write and the design teams can therefore focus only on the desired behavior;

– facilitating behavioral intellectual property reuse and retargeting [CON 11]. In addition to R&D productivity improvement, behavioral synthesis has the added value of allowing efficient reuse of behavioral IPs. Meanwhile, compared with RTL IPs that have fixed microarchitecture and interface protocols, behavioral IPs can be retargeted to different implementation technologies or system requirements;

– adding to the maintainability and facilitating the updating of the production. Huge silicon capacity may require a large code density for behavior description, which is impossible to be easily handled by human designers for product maintenance or updating. Higher levels of abstraction are one of the most effective methods to reduce the code density [WAK 05].

Xilinx [XIL 13b] reports that its HLS-incorporated product, Vivado_HLS, can accelerate design cycles by around 30% with significant performance benefits related to the conventional methods (see Figure 4.2). Meanwhile, the case of Wakabayashi [WAK 05] shows that a 1M-gate design usually requires about 300 K lines of RTL code, which cannot be easily handled manually, while the code density can be easily reduced by 7–10x when moved to high-level specification in C-like languages, resulting in a much reduced design complexity. Likewise Villarreal *et al.* [VIL 10] present a riverside optimizing compiler for configurable circuits that achieves an average improvement of 15% in terms of the metrics of lines of code and programming time over handwritten VHDL in evaluation experiments.

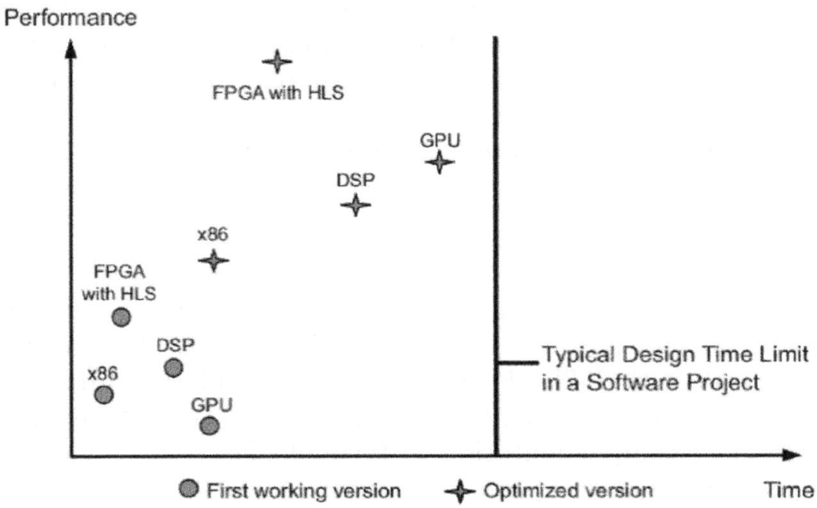

Figure 4.2. *Design time versus application performance with Vivado_HLS compiler*

HLS design flows have significant values. The electronic manufacturers constantly strive to improve their productivity by raising the abstraction level and easing the design process for their engineers due to the increasingly furious market competition worldwide. After an effort of over 20 years, many high-quality HLS tools have been made available for practical applications [MEE 12, PRO 14, RUP 11], i.e. Vivado_HLS of Xilinx [XIL 12b] and Catapult C Synthesis Work Flow [WAN 10]. According to the studies of Meeus *et al.* [MEE 12], there are five commonly used HLS tools, including Vivado_HLS (Auto Pilot) [ZHA 08], CatapultC [WAN 10], C-to-Silicon [CAD

11], Cyber Workbench [WAK 05] and Synphony C [SYN 10]. Table 4.1 compares their characteristics by using the Sobel edge detection algorithm from eight aspects. Evaluation results are represented by using five levels, --, -, +-, + and ++ from low to high, respectively. The abstraction level evaluates the ability of the tools in terms of timing control. Data types refer to the capability for data format support, such as floating or fixed-point data. Exploration is the design space exploration ability of the tool. Verification evaluates whether the tool can help to generate easy-to-use test benches quickly. RTL quality is estimated by the amount of the hardware consumption of the generated RTL. Documentation is used to evaluate whether the tool is easy to learn or if its documentation is extensive or not. Ease of implementation indicates whether the original source code could be used with fewer modifications or needs to be rewrite completely.

	Auto Pilon Vivado_HLS	Catapult C	C-to-silicon	Cyber WorkBench	Synphony C
Source	C/C++ System C	C/C++ System C	C, TLM System C	C, BDL System C	C/C++
Abstraction level	++	++	+	++	++
Data type	+	+	+	+	+
Design exploration	++	++	-	++	+-
Verification	++	++	+	+	++
RTL quality	++	+	+-	++	+-
Documentation	++	++	+	+-	+
Learning curve	++	++	--	+	+
Ease of implementation	++	++	+	+	++

Table 4.1. *Characteristic evaluation of five HLS tools*

We find that Vivado_HLS has a near perfect performance related to the others. It can benefit the desired design suite in the following aspects:

– scheduling untimed code (operations) without any restrictions;

– handling floating-point variables by mapping them onto fixed points: this facilitates the designs of high accuracy applications;

– providing extensive and intuitive exploration options: with Vivado_HLS, users can discover new solution alternatives by reconfiguring the optimization directives instead of modifying the source code;

– ability to estimate the running cost of the design: this can facilitate the evaluation of the design candidates and making the final decision;

– generating high-quality RTL implementations: designers always strive for better design performances within their powers. According to the measuring of Meeus *et al.* [MEE 12], Vivado_HLS could save up to and more than 95% of hardware resources related to the other 11 HLS tools;

– easy to be learned even for those who are less knowledgable about FPGAs: this tool can be quickly mastered by C users and requires less hardware knowledge. Furthermore, its documentation is extensive and easy to understand.

4.2. Vivado_HLS process presentation

Vivado_HLS can synthesize the desired algorithm specified using the HLS-available C language into RTL and performs the RTL ports for the top behavior with specific timing protocol depending on the function arguments. Generally, its overall process consists of control and datapath extraction, scheduling and binding, data type processing, optimizations and design constraints. This section discusses the two most important cycles of synthesis process of HLS: control and datapath extraction and scheduling and binding, which may offer most benefits of performance optimization.

4.2.1. *Control and datapath extraction*

C is a structured programming language that consists of sequential, selectional and repetitional statements, so Vivado_HLS begins the synthesis process with the control and datapath extraction including control behavior extraction, operations extraction and control and datapath creation.

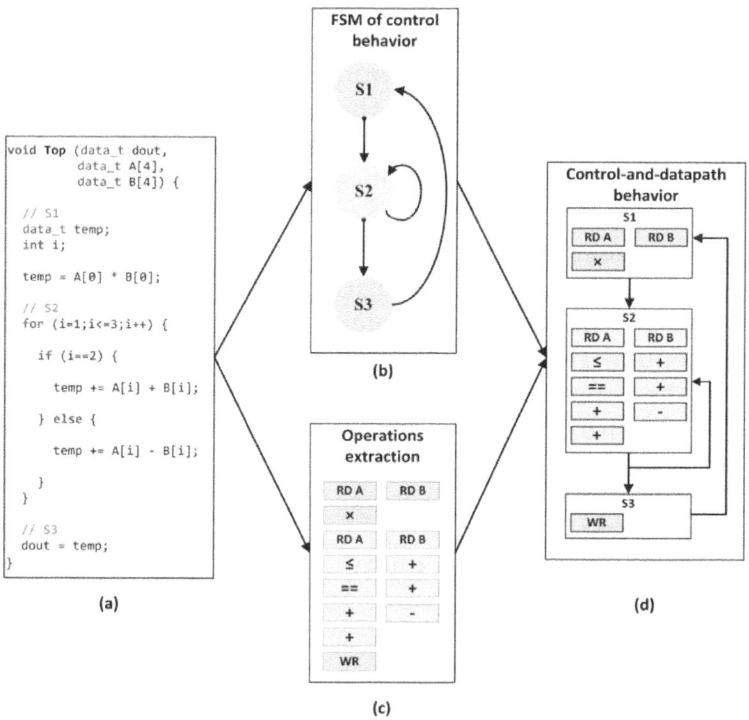

Figure 4.3. *HLS control and datapath extraction example: a) input C code source, b) control extraction, c) operation extraction and d) generated control and datapath behavior. For a color version of the figure, see www.iste.co.uk/li/image.zip*

First of all, Vivado_HLS analyzes the control logic of the source code to extract its control behavior as a finite-state machine (FSM). For example, the program shown in Figure 4.3(a) is cut into two blocks *S1* and *S3* by the loop *S2*. The content of *S1* consists of variable declarations and a multiplication operation, *S2* is a loop whose body is an *If* statement and *S3* contains only an assignment statement. We assume that all the state transitions can finish their jobs (operations) in a single cycle. Hence, the control behavior FSM for this example should be as shown in Figure 4.3(b).

Next, all the operations in the source code are extracted. Figure 4.3(c) lists all the operations extracted from (a) and classifies them according to the states by the colors of red, green and blue. Since the operation extraction does not take into account register assignments, *S1* has only three operations including reading the elements to calculate from array A and B and finding their product. Relatively, the treatment of *S2* is more complicated because it does not only have to finish the operations required by the algorithm but also needs some extra operations due to the control logic. Moreover, although each iteration follows only one of the branches in the *if-else* statement, both of them need to be implemented. Hence, the extracted operations of *S2* include a <= and a + for the loop, a == for *if-else* and three + and a - for the algorithm. *S3* assigns the final result stored in *temp* to the argument *dout*, so it has only a writing operation. Finally, the extracted operations are assigned to the control behavior FSM and perform the control and datapath behavior shown in Figure 4.3(d).

4.2.2. *Scheduling and binding*

Scheduling and binding is the key aspect of HLS. The scheduling process determines in which cycles the operations occur, while binding determines which hardware resource or core is used. Figure 4.4 shows the scheduling and binding flow. During this process, the extracted control and datapath behavior and multiple constraints from the device technique libraries and user configurations have to be taken into account, such as the clock frequency, clock uncertainty, timing information, area, latency and throughput directives, etc.

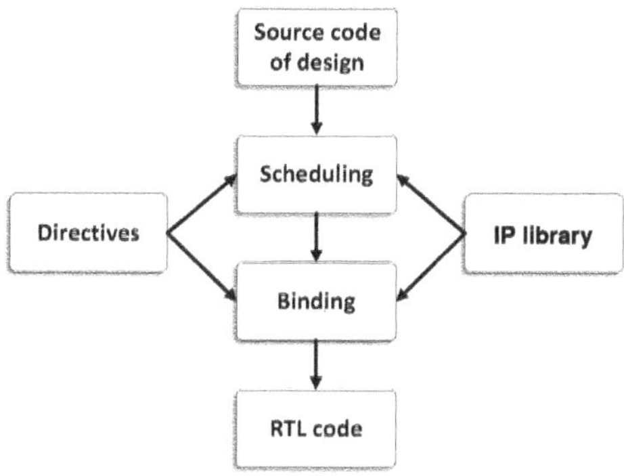

Figure 4.4. *HLS scheduling and binding flow*

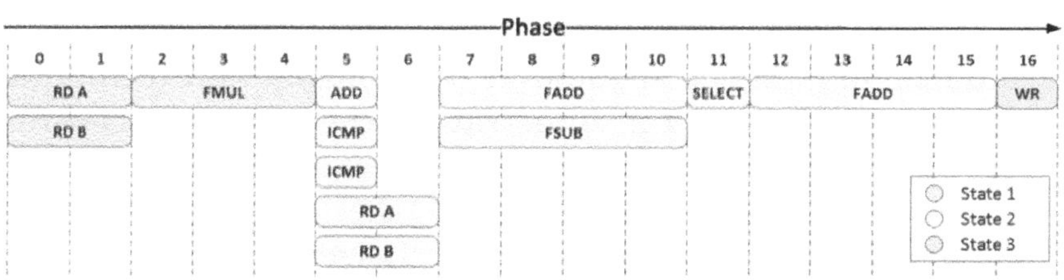

Figure 4.5. *Scheduling of the example in Figure 4.3(a). For a color version of the figure, see www.iste.co.uk/li/image.zip*

Let us consider Figure 4.3(a) for instance, in which we define the data type as float and two independent interfaces are implemented for Vector A and Vector B. Its default scheduling result with *xc7vx1140tg1928-1* of Xilinx is shown in Figure 4.5. In this schematic, a cycle equivalents to a transition from Phase N to Phase $N + 1$, i.e. *RD* consumes a cycle because it takes up two phases while *ADD* does not consume any cycles as it takes only a single phase. Due to the independent interfaces, the *RD* operations of each vector can be parallelized, so they simultaneously occur in the same phase. According to its control and datapath behavior, *S2* is a loop associated with an *if-else* structure. Generally, the control operations, including *ADD*, *ICMP* and *SELECT*, do not consume cycles, therefore they are usually parallelized with other operations if the sequential relationship allows, i.e. *RD A* and *RD B* in Phase 5. Once the comparator in Phase 5 finds that the next iteration will be beyond the loop boundary, it breaks the loop and goes to the next state immediately. In the body of the loop, the two branches of the *if* statement occur in the same cycles but only one of them is selected via the selector in Phase 11. That is, only one of the two operations is activated in each iteration in the final RTL implementation.

After the scheduling process, the binding process maps the operations to the cores in the technique library. There are generally two binding methods: sharing component and non-sharing component. HLS determines which strategy to use depending on the sequential relationship and technique constraints. Table 4.2 illustrates the operations-cores mapping for the scheduling schematic of Figure 4.5. In this case, *FADD* and *FSUB* use the same IP core: *faddfsub_32ns_32ns_32_4_full_dsp*. Because there are no simultaneously running *FADD/FSUB* operations in the same cycles, HLS makes them share a single floating-point adder U1. On the other hand, the multiplication operation appears once in the process, therefore only a single multiplier is implemented.

Operations	IP Cores	Units
FADD	*faddfsub_32ns_32ns_32_4_full_dsp*	U1
FSUB	*faddfsub_32ns_32ns_32_4_full_dsp*	U1
FMUL	*fmul_32ns_32ns_32_3_max_dsp*	U2

Table 4.2. *Operations-cores mapping of the scheduling schematic in Figure 4.5*

In this section, we can see that the HLS process is highly available to automate the C-to-RTL conversion. Now the question is coming, can the HLS generated RTLs satisfy the industrial level requirements in terms of performance in real-life applications? For this issue, we implement a skin lesion assessment system by using it to evaluate its performance.

4.3. Case of HLS application: FPGA implementation of an improved skin lesion assessment method

The newly introduced ASCLEPIOS [JOL 11] is a promising multispectral imaging system for the assessment of skin lesions (see Figure 4.6). It provides images of skin reflectance at several spectral bands (visible + near infrared spectrum), coupled with software that reconstructs a reflectance cube from the acquired images. The reconstruction is performed by neural network based algorithms. This yields an increase in the spectral resolution without the need for an increase in the number of filters. Reflectance cubes are generated to provide spectral analyses of skin data that may reveal certain spectral properties or attributes not initially obvious in multispectral images.

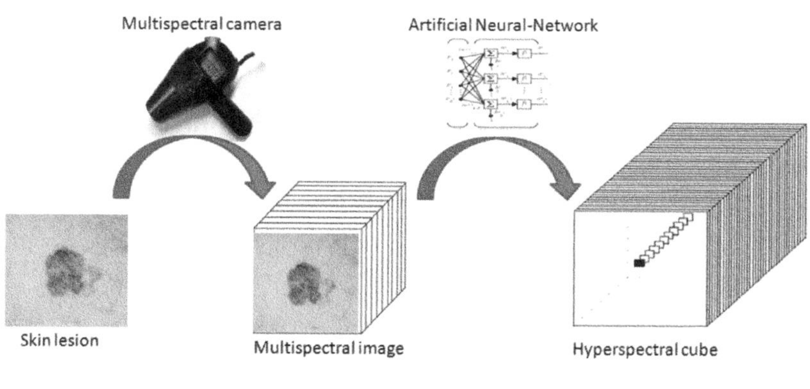

Figure 4.6. *ASCLEPIOS system illustration. For a color version of the figure, see www.iste.co.uk/li/image.zip*

Basing on such a system, a KMGA is developed to retrieve the skin parameter maps from the reflectance cube. Using five skin parameter maps such as concentration or epidermis/dermis thickness, this method combines the Kubelka–Munk light-tissue interaction model and genetic algorithm function optimization process to produce a quantitative measure of cutaneous tissue. In this HLS evaluation example, we realize the FPGA implementation of an improved version of the KMGA. During the development, several optimizations are made in order to improve the performances of the final RTL implementation, including optical function rewriting, function optimizer improving and memory optimizing. Obtained experiments demonstrate the high feasibility of low-cost and high-practicability hardware equipment for the clinical system.

4.3.1. *KMGA method description*

4.3.1.1. *Light propagation model in the skin*

In order to retrieve the different physical or biological skin properties, several skin models have been developed. The Kubelka–Munk model [KUB 31] is one of the most popular and simplest approaches for computing light transport in a highly scattering medium and has been widely used to model the light-skin interaction. The skin is seen as a 2-layer medium (epidermis and dermis) with five principal parameters that affect the reflectance and transmittance: melanin concentration, epidermis thickness, blood concentration, blood oxygen saturation and dermis thickness. The total reflectance R_{tot} and transmittance T_{tot} are expressed as:

$$R_{tot} = R_{1,2} = R_1 + \frac{T_1^2 R_2}{1 - R_1 R_2} \qquad [4.1]$$

$$T_{tot} = T_{1,2} = \frac{T_1 T_2}{1 - R_1 R_2} \qquad [4.2]$$

The reflectance R_n and transmittance T_n for a single layer n can be expressed as a function of the thickness of the layer d_n, the absorption coefficient $\mu_{a,n}$ and the scattering coefficient $\mu_{s,n}$:

$$R_n = \frac{[1-\beta_n{}^2]\times[exp(K_n d_n)-exp\,(-K_n d_n)]}{[1+\beta_n]^2 exp(K_n d_n)-[1-\beta_n]^2 exp\,(-K_n d_n)} \quad [4.3]$$

$$T_n = \frac{4\beta_n}{[1+\beta_n]^2 exp\,(K_n d_n)-[1-\beta_n]^2 exp\,(-K_n d_n)} \quad [4.4]$$

where:

$$K_n = \sqrt{\mu_{a,n}(\mu_{a,n} + 2 \times \mu_{s,n})} \quad [4.5]$$

$$\beta_n = \sqrt{\frac{\mu_{a,n}}{\mu_{a,n}+2\mu_{s,n}}} \quad [4.6]$$

The suffix n equals 1 or 2 for epidermis or dermis. The optical absorption coefficient of epidermis layer $\mu_{a.epidermis}$ is known as a function of concentration of melanin f_{mel}, the melanin spectral absorption coefficient $\mu_{a.melanosome}$ and the baseline skin absorption coefficient $\mu_{a.baseline}$:

$$\mu_{a.epidermis} = f_{mel}\,\mu_{a.melanosome} + (1 - f_{mel})\,\mu_{a.baseline} \quad [4.7]$$

where, with λ corresponding to wavelength parameter,

$$\mu_{a.melanosome} = 6.6 \times 10^{11} \lambda^{-3.33} \quad [4.8]$$

$$\mu_{a.baseline} = 0.244 + 85.3 \times exp\frac{-(\lambda-164)}{66.2} \quad [4.9]$$

On the other hand, the dermal absorption coefficient $\mu_{a.dermis}$ is expressed as follow:

$$\mu_{a.dermis} = f_{blood}\left(C_{oxy}\,\mu_{a.oxy}\right) + f_{blood}\left(1 - C_{oxy}\right)\mu_{a.deoxy} + (1 - f_{blood})\,\mu_{a.baseline} \quad [4.10]$$

where f_{blood} is the blood concentration in %, C_{oxy} is the oxygen saturation in blood, and $\mu_{a.oxy}$ and $\mu_{a.deoxy}$ are the absorption coefficient of the oxy-hemoglobin and deoxy-hemoglobin in cm^{-1}, respectively. The values of $\mu_{a.oxy}$ and $\mu_{a.deoxy}$ with the different wavelengths can be calculated from the Takatani–Graham table using the following equations [JAC 08]:

$$\mu_{a.oxy} = ln\,10 \times HbO_2(\lambda) \times G/M \quad [4.11]$$

$$\mu_{a.deoxy} = ln\,10 \times Hb(\lambda) \times \frac{G}{M} \quad [4.12]$$

where HbO_2 and Hb are the oxy-hemoglobin and deoxy-hemoglobin content in cm^{-1}/M, G is the hemoglobin's weight in gram per liter and M is the gram molecular weight of hemoglobin. We selected 150 g/L and 64,500 g/mol as G and M [TAK 79]].

The scattering coefficients of both layers $\mu_{s,epidermis}$ and $\mu_{s,dermis}$ are the sum of the Mie scattering coefficient $\mu_{s,Mie}$ and the Rayleigh coefficient $\mu_{s,Raylight}$:

$$\mu_{s.epidermis} = \mu_{s.dermis} = \mu_{s.Mie} + \mu_{s.Raylight} \quad [4.13]$$

where

$$\mu_{s.Mie} = 2 \times 10^5 \times \lambda^{-1.5} \quad [4.14]$$

$$\mu_{s.Raylight} = 2 \times 10^{12} \times \lambda^{-4} \quad [4.15]$$

4.3.1.2. *Function optimizer for skin optical properties retrieval*

According to previous equations, it is possible to express $R(\lambda)$, the total reflectance of the incident light with a certain wavelength, as a function of the skin parameters:

$$R\,(\lambda) = f_{KM}\left(f_{mel}, D_{epi}, f_{blood}, C_{oxy}, D_{dermis}\right) \quad [4.16]$$

where D_{epi} and D_{dermis} are the thickness of the epidermis layer and the dermis layer.

The reflectance values are measured with an acquisition system. After a preprocessing step, we can obtain a reflectance cube with two spatial dimensions and one spectral dimension. This cube can be seen as an image where each pixel is described by a vector, representing the reflectance spectrum of the skin as a function of the wavelength (see Figure 4.7). From this measure, the objective is to find the five skin parameters for each pixel that are the most suitable for this measured reflectance spectrum. It is obvious that the KM model is a nonlinear function with five arguments. Several studies have been done about how to solve the similar complex nonlinear function, and the findings of Viator *et al.* [VIA 05] and Choi [CHO 10] have proved that genetic algorithms (GA) are an effective approach.

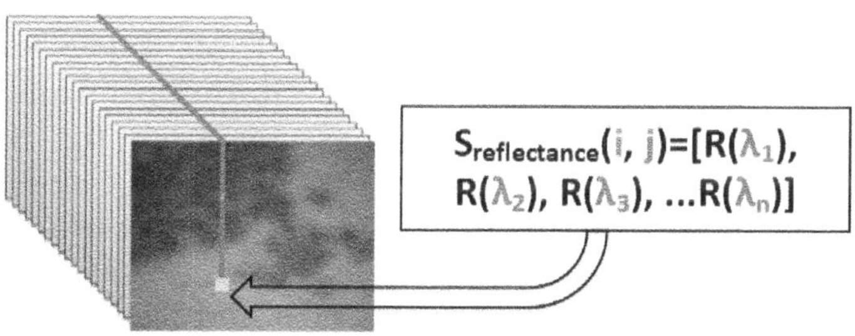

Figure 4.7. *Reflectance spectrums $S_{reflectance}$ at a single pixel formed from the reflectance measured: blue and red represent the pixel's position and green represents the different wavelength values. For a color version of the figure, see www.iste.co.uk/li/image.zip*

In KMGA, a candidate solution (called *individual*) is composed of the reflectance spectrum, skin parameters (f_{mel}, D_{epi}, f_{blood}, C_{oxy}, D_{dermis}), the spectrum generated by these parameters using equation [4.16] and a fitness value as shown in Figure 4.8. This last one depends on the similarity between the spectrum generated by the parameters and the measured spectrum. It can be expressed by a metric scale such as root mean squared error (RMSE). The set of individuals is called a *population*.

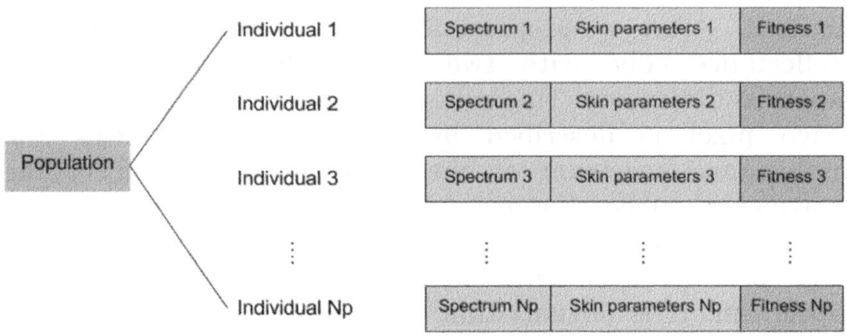

Figure 4.8. *Population data structure: the skin parameters are f_{mel}, D_{epi}, f_{blood}, C_{oxy} and D_{dermis}. For a color version of the figure, see www.iste.co.uk/li/image.zip*

Figure 4.9 illustrates the overall implementation of GA for the KM inversion. First, N_P individuals are randomly generated within the range shown in Table 4.3 to form an initial population. Then, in an iterative framework, population evolves by techniques inspired by natural evolution. The evolution of the population is composed of three steps: best individual selection, crossover-mutation and random selection. During the selection process, only the best individuals are kept for the next iteration. Then, the crossover process selects multiple couples of individuals (parents) from the population and one crossover parameter from the five skin parameters to create two new individuals (offspring) by swapping the parents' selected parameter values. Finally, the mutation process randomly generates some new skin parameter values to replace certain individuals' old ones. The aim of mutation is to keep exploring the search space and avoid being trapped into a local minimum. These processes are repeated until a predefined number of iterations. Finally, the best candidate is selected.

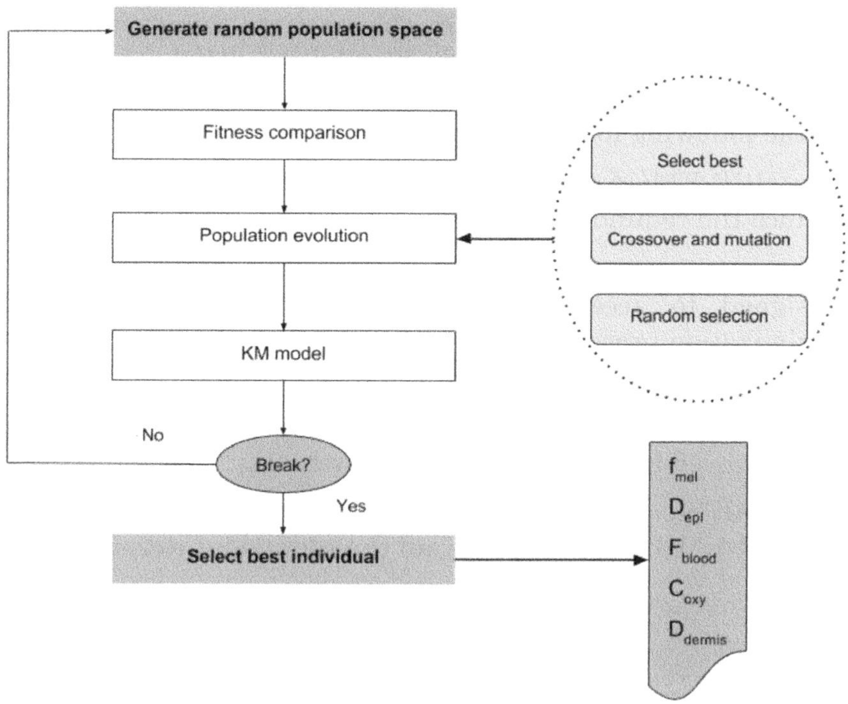

Figure 4.9. *Overall genetic algorithm procedure for KM model inversion*

Skin parameter	Symbol	Range
Melanin concentration	f_{mel}	1.3–43%
Epidermis thickness	D_{epi}	0.01–0.15 mm
Volume blood fraction	f_{blood}	0.2–7%
Oxygen saturation	C_{oxy}	25–90%
Dermis thickness	D_{derm}	0.6–3 mm

Table 4.3. *Size of search spaces for skin parameters (see [JOL 13])*

4.3.2. *KMGA method optimization*

KMGA could effectively retrieve the skin parameter maps via a selection process; however, this task is very time consuming even for a powerful processor. Thus, we study a novel HCR-KMGA version in this section. Comparing with its prototype implemented by Jolivot [JOL 11], this implementation can make more acceleration gains according to the following four approaches:

– HCR-KMGA respecifies the KM function in order to reduce the redundant operations down to a minimum;

– a prediction function optimization algorithm (PFOA) is designed to accelerate the convergence of the function optimization process;

– HCR-KMGA individuals' parameters are optimized depending on the data dependency; some unnecessary data are removed in order to save memory space;

– multiple different termination conditions are performed in HCR-KMGA in order to avoid the redundant iterations.

4.3.2.1. *Function reformulation for computation complexity reduction*

According to Figure 4.9, the KMGA method consists mainly of population initialization, population generation and population evolution. Experimental results show that the population initialization and generation takes up to 96% of the total execution time, in which population evolution takes 3% and other operations only 1%. KM is the key technique used during the time-consuming process of population initialization and generation. Thus, in order to speed up the KMGA method, we definitely have to improve its efficiency.

The KM function, the set of equations [4.3]–[4.6], shows that the original model is complex and most parts of its instructions are costly and redundant. For example, the $exp(K_n d_n)$ term appears four times in equation [4.3]. These redundant instructions are insignificant and costly. It is possible to avoid them by arithmetical reduction. Thus, we reform the KM model:

$$R_n = F_R(f_{mel}, D_{epi}, f_{blood}, C_{oxy}, D_{dermis}) \quad [4.17]$$

$$T_n = F_T(f_{mel}, D_{epi}, f_{blood}, C_{oxy}, D_{dermis}) \quad [4.18]$$

where F_R and F_T are optimized KM model functions. These symbols are expressed by:

$$F_R = \frac{\mu_{s,n} \times (E-1)}{(\mu_p + K_n) \times E - (\mu_p - K_n)} \quad [4.19]$$

$$F_T = \frac{2K_n \varepsilon}{(\mu_p + K_n) \times E - (\mu_p - K_n)} \quad [4.20]$$

where

$$\mu_p = \mu_{a,n} + \mu_{s,n} \quad [4.21]$$

$$E = \varepsilon^2 = e^{2K_n d_n} \quad [4.22]$$

The above rewritten KM functions are simple and effective at keeping the processors' obligations down to a minimum because reading the calculated data from a local register, instead of repeating the computations, can provide a great gain in executive time. Here, *exp()* is the most costly function, therefore we cut its repetitions down to only once per iteration by combining like terms and defining two interim storing registers E and ε. The other instructions are also reduced more or less by similar approaches. Table 4.4 compares the number of necessary instructions between the initial prototype and our implementation: the optimized function requires fewer instructions than before.

Model	Addition/ subtraction	Multiplication/ division	Exponentiation
Original KM	13	17	13
Optimized KM	8	9	3

Table 4.4. *Necessary instruction number comparison between original and optimized KM functions*

On the other hand, section 4.3.1 presents that the optical absorption and scattering coefficients of epidermal and dermal layers are ultimately functions of wavelength and skin parameters. We can reformulate equations [4.7], [4.10] and [4.13] by using vector equations:

$$U_{a.epidermis} = f_{mel}\, U_{a.melanosome} + (1 - f_{mel})\, U_{a.baseline} \quad [4.23]$$

$$U_{a.dermis} = f_{blood}\left(C_{oxy}\, U_{a.oxy}\right) + f_{blood}\left(1 - C_{oxy}\right) U_{a.deoxy} + (1 - f_{blood})\, U_{a.baseline} \quad [4.24]$$

$$U_{s.epidermis} = U_{s.dermis} = U_{s.Mie} + U_{s.Raylight} \quad [4.25]$$

where

$$U_{a.melanosome} = 6.6 \times 10^{11} \times \begin{bmatrix} \lambda_1^{-3.33} \\ \lambda_2^{-3.33} \\ \lambda_3^{-3.33} \\ \vdots \\ \lambda_n^{-3.33} \end{bmatrix} \quad [4.26]$$

$$U_{a.baseline} = 0.244 + 85.3 \times \exp\left\{-\left(\begin{bmatrix} \lambda_1 \\ \lambda_2 \\ \lambda_3 \\ \vdots \\ \lambda_n \end{bmatrix} - 164\right)/66.2\right\} \quad [4.27]$$

$$U_{a.oxy} = ln\,10 \times \begin{bmatrix} HbO_2(\lambda_1) \\ HbO_2(\lambda_2) \\ HbO_2(\lambda_3) \\ \vdots \\ HbO_2(\lambda_n) \end{bmatrix} \times G/M \quad [4.28]$$

$$U_{a.deoxy} = ln\,10 \times \begin{bmatrix} Hb(\lambda_1) \\ Hb(\lambda_2) \\ Hb(\lambda_3) \\ \vdots \\ Hb(\lambda_n) \end{bmatrix} \times G/M \quad [4.29]$$

$$U_{a.Mie} = 2 \times 10^5 \times \begin{bmatrix} \lambda_1^{-1.5} \\ \lambda_2^{-1.5} \\ \lambda_3^{-1.5} \\ \vdots \\ \lambda_n^{-1.5} \end{bmatrix} \quad [4.30]$$

$$U_{a.Rayleight} = 2 \times 10^{12} \times \begin{bmatrix} \lambda_1^{-4} \\ \lambda_2^{-4} \\ \lambda_3^{-4} \\ \vdots \\ \lambda_n^{-4} \end{bmatrix} \quad [4.31]$$

In equations [4.23]–[4.31], $U_{a.melanosome}$, $U_{a.baseline}$, $U_{a.oxy}$, $U_{a.deoxy}$, $U_{a.epidermis}$ and $U_{a.dermis}$ are six coefficient vectors with different wavelengths. They are fixed from 450 to 780 nm with a step of 10 nm. These vectors can be precalculated and stored in the memory. Thus, instead of calculating the values wanted at each iteration, the processor is able to read them directly from the memory. This approach avoids all the repeated computations due to equations [4.8], [4.9] and [4.11]–[4.15] in the routine.

4.3.2.2. *Convergence acceleration of function optimizer*

It is obvious that equation [4.16] is a complex nonlinear function with five arguments, which is impossible to inverse by mathematical methods. KMGA optimizes this function according to a standard genetic algorithm. This optimization process is a search heuristic that mimics the process of natural selection. It generates solutions to optimization problems using techniques inspired by natural evolution, such as inheritance, mutation, selection and crossover. It has also proved effective in the application of bioinformatics, phylogenetics, computational science, engineering, economics, chemistry, manufacturing, mathematics, physics and pharma-cosmetics, etc. However, the evolution process of a pure natural-simulated genetic algorithm is time consuming, and can easily get trapped into local optima. This is because GA always generate the new populations in a random way first and then selects the best individual according to the fitness function, which enormously reduces the chance to find a better individual in the next iteration that results in a very low convergence rate. Thus, within

HCR-KMGA, we perform a PFOA optimization process, which can raise the convergence rate by predicting the possible evolution directions.

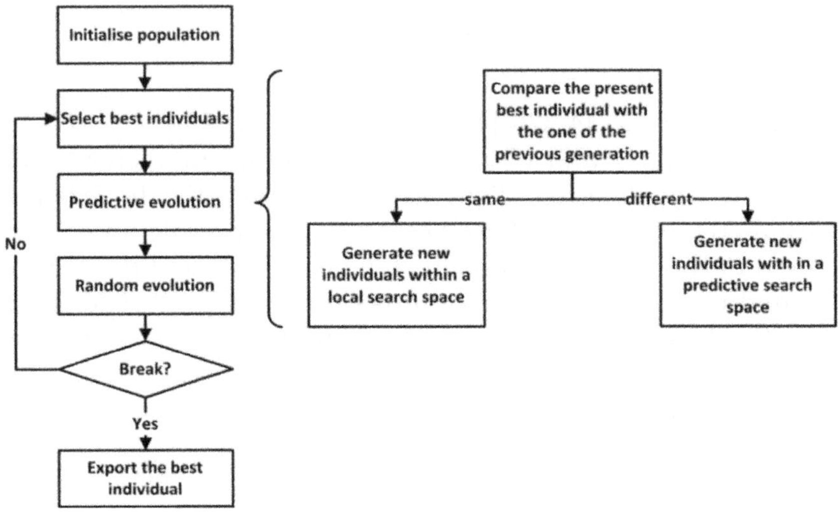

Figure 4.10. *Overall architecture of the prediction function optimization algorithm (PFOA)*

Figure 4.10 illustrates the overall architecture of PFOA. Like the conventional GA, the system first randomly initializes the population. However, in the evolution process, only the best individual selection process is kept, while crossover-mutation and random selection are replaced by predictive evolution and random evolution. After each iteration, the best individuals are directly copied from the last generation into the next one for the purpose of fast convergence. Meanwhile, some of the individuals move depending on a prediction strategy, which can further raise the convergence rate of population evolution by a considerable amount. Finally, the rest of the individuals are re-performed randomly in order to reduce the possibility of falling down to the local optima.

Depending on different fitness functions, designers can customize different prediction strategies. In our case, an individual has five skin tissue parameters and the size of population consists of a few hundred individuals. In each iteration, several new genes are generated via the crossover-mutation process. However, with the floating-point number, which is one of the most frequently used data formats in computer science, KMGA has more than 6E27 candidates (this value is estimated according to the range of the five skin parameters shown in Table 4.3 and the precision of the floating-point numbers), which may result in a long running time and a very low convergence rate.

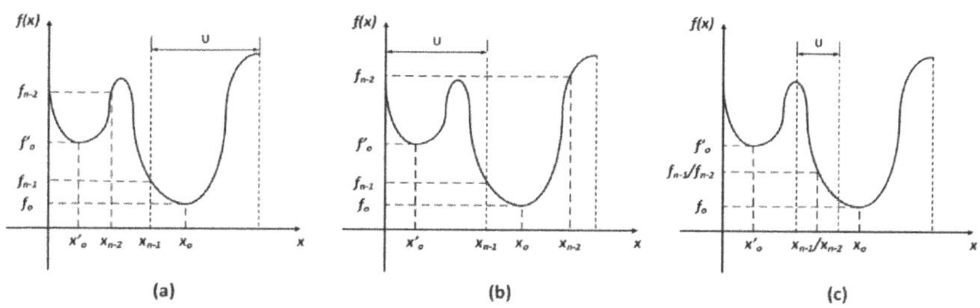

Figure 4.11. *Search space prediction of PFOA: x_n and f_n are the parameter and fitness value of the nth iteration's best individual, (x_o', f_o') is the local optima and (x_o, f_o) is the global optima of the optimization function*

For the purpose of accelerating the convergence speed of the algorithm, a prediction strategy that can reduce each iteration's search space by predicting the evolution direction is performed as shown in the right of Figure 4.10. First, the best individuals of the last two generations are compared and, depending on the comparison result, the algorithm takes different steps to adjust the search space. We base the prediction strategy on the assumption that higher parameter values had better fitness while $x_{n\text{-}1} > x_{n\text{-}2}$, and lower parameter values had better fitness while $x_{n\text{-}1} < x_{n\text{-}2}$, where x_n refers to the parameter value of the best individual of the nth generation. As shown in Figures 4.11(a) and (b), PFOA locks the search space U onto the scope of $x > x_{n-1}$ with $x_{n-1} > x_{n-2}$, while the scope of $x < x_{n-1}$ with $x_{n-1} < x_{n-2}$. Meanwhile, we note that this method is effective only for the situations in which the present best individual is far enough away from the optima. Once it is closer to the optima, a much smaller search space may be required to enable the algorithm to find a better individual with as few iterations as possible. This situation is abstracted as $x_{n-1} = x_{n-2}$ and $f(x_{n-1}) = f(x_{n-2})$ in PFOA. It means that no better individuals are found in the last two iterations. Therefore, the search space of the nth iteration will be locked within the scope around x_{n-1} in order to enhance the chance of evolution (see Figure 4.11(c)).

Since the optimization function is unknown, it is impossible to always correctly predict the position of the global optima. However, this mistake can be quickly corrected in the following iterations. For example, the predicting scope does not include the optima in Figure 4.11(b) and within this scope no better individual can be found. However, this makes the algorithm restrict its search space around x_{n-1} in the following iterations, within which a new best individual can be easily found in the right of x_{n-1}.

The KM model has five parameters to be figured out. Considering that these parameters have different effects on the final fitness, their analyses must be independent from each other. Thus, HCR-KMGA applies PFOA to all of them respectively. That is, after the fitness comparison, the search space of each parameter is defined independently via the proposed prediction strategy.

It should also be noted that sometimes this method may also lead the evolution down to local optima. Thus, after prediction evolution, some random individuals are performed in order to avoid this outcome. Unlike GA, PFOA completely regenerates all the individuals in a random way instead of crossovers or mutations. This method greatly enriches the sample types of genes, so the risk of missing the optima is reduced.

4.3.2.3. *Individual information storage optimization*

Figure 4.8 shows that the information of an individual consists of fitness value, chemical properties (skin parameters) and optical properties (simulated spectrum). These data need to be saved in the memory of the processing device permanently throughout the whole processing, and therefore the conventional KMGA prototype has to consume a lot of hardware resources to store all the population information, especially for embedded devices. Thus, an approach that can reduce memory consumption is required.

According to the algorithm analysis, it is found that each piece of individual information is not isolated but rather has internal relations with the others. Figure 4.12 displays the relationships among them and it demonstrates that both the simulated optical properties and the fitness value are calculated from the chemical properties. That provides a nice opportunity to compress the data size by removing the data that could be computed immediately and keeping only the necessaries.

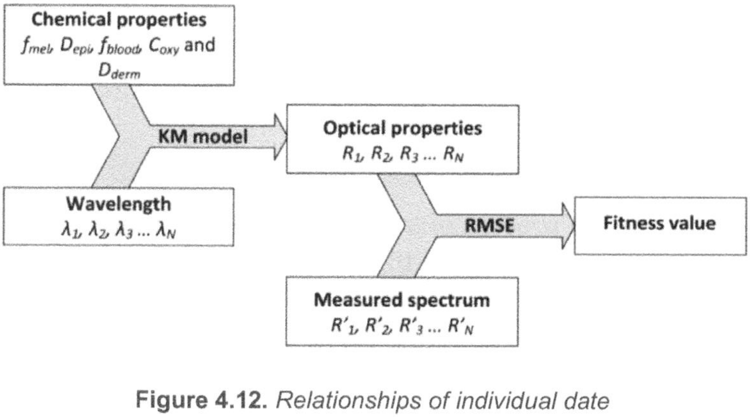

Figure 4.12. *Relationships of individual date*

The individuals of HCR-KMGA are performed by chemical properties and fitness values, while the information of its simulated optical properties is removed. In the overall algorithm, the simulated spectrum of the skin lesion zone is used only once in order to compute the fitness values for the best individual selection at the beginning of each iteration (see Figure 4.10) and it is therefore unnecessary to allocate a certain memory to store them. Meanwhile, a single fitness value takes up only several octets while its computing has to call the KM models, which has a very long running time. Nevertheless, this data is required not only in the best individual selection process but also in the prediction evolution process. Thus, this number is stored as part of the individual information in HCR-KMGA in order to accelerate the design by reducing the operations number. The data size comparison between KMGA's and HCR-KMGA's individual could be expressed as follow:

$$\frac{L_{KMGA}}{L_{HCR-KMGA}} = \frac{6 + N_{spectrum}}{6} \quad [4.32]$$

where L_{KMGA} and $L_{HCR\text{-}KMGA}$ are the total data length of the two algorithms' individuals and $N_{spectrum}$ is the number of bands of the spectral image. Obviously, HCR-KMGA consumes much less memory space than KMGA and its total length is permanently six times the defined number length. That is, the algorithm will not take any more hardware resources for the population information storage even for spectral images with high bands numbers.

4.3.2.4. *Iteration termination conditions*

The evolution process of a GA is terminated after a number of iterations according to the termination conditions customized by designers. A main issue that always affects the selection of termination conditions is that defining a condition which is easy to reach consumes fewer hardware resources but may reduce the accuracy performance of designs, while a hard condition may lead to extensive computational time and the algorithm can even be trapped in an infinite loop.

KMGA terminates the evolution process by defining an iteration number which corresponds to the convergence of the population. That is, the evolution stops when the iterations number reaches the default value. This approach could provide an acceptable average fitness value for the processing of a standard skin lesion multi-spectral image. However, the computation of GA is full of all kinds of possibilities, so such a simple termination condition may lead to redundant iterations or unpredictable results. For example, the evolution process will not be terminated until it reaches the default iterations number even if the global optima has been found, or alternatively the evolution process may have been terminated before the fitness reaches the required level. Thus, instead of forcing the algorithm to end by setting a default iteration limitation, HCR-KMGA combines multiple termination conditions together, including max continuous invalid iteration level $\theta_{invalid.iter}$, fitness level $\theta_{fitness}$ and total iteration level $\theta_{total.iter}$.

In our work, an iteration in which a new best individual is found is called a valid iteration, otherwise it is an invalid one. Once a large number of continuous invalid iterations appear, it means that the present best individual is very close to the optima and it will be difficult or time-consuming to find another one for the algorithm. So in order to save the hardware resources, a max continuous invalid iteration level is defined to break the evolution loop while no new best individual is found during $\theta_{invalid.iter}$ iterations.

Normally, the goal of evolution is not to exactly figure out the optima. That is, a fitness error could be accepted in each processing. In HCR-KMGA, a fitness level refers to the acceptable fitness error. Once the fitness value of the present best individual is lower than the fitness level, the evolution loop could be broken as well.

Finally, in order to avoid the algorithm getting trapped in an infinite loop, a total iteration level is defined. It should be noted that these three termination conditions simultaneously effect the algorithm, so no matter which one is reached, the iterations will be ended. This method can save as many hardware resources as possible, and meet the accuracy requirements of the design as well.

4.3.3. *HCR-KMGA implementation onto FPGA using HLS technique*

As one of the most popular computation platforms, FPGAs have been used in a wide variety of real-time image processing applications. Depending on the proposed HCR-KMGA, a FPGA-based embedded skin lesion assessment system is realized in this section.

4.3.3.1. *Development flow*

Figure 4.13 illustrates the HLS-based development process used for the implementation. After a careful algorithm analysis, we first realize a synthesizable KMGA prototype in C language. Next, the prototype is optimized by using the approaches discussed in section 4.3.2 and simulated through a common C compiler, i.e. Intel C++ Compiler (ICC) [INT 08]. Third, the functionally verified source code is imported into the HLS tool for C-to-RTL synthesis. During this process, the synthesis directives are carefully configured in order to effectively use the hardware resources of the target device. Finally, the IP block of HCR-KMGA is automatically performed for top behavior.

4.3.3.2. *Functional block implementation onto FPGA device*

The over-all architecture of HCR-KMGA is shown in Figure 4.14. Its initial population is generated according to the skin parameters' value range presented in Table 4.3. As mentioned in section 4.3.2, we replace the calculation of $\mu_{a.mel}$, $\mu_{a.baseline}$, $\mu_{a.oxy}$, $\mu_{a.deoxy}$ and $\mu_{s.epidermis}$ with different wavelengths by a coefficient table pre-calculated using the Takatani–Graham table [TAK 79]. As there is no golden rule to define the values of evolution process parameters, we base HCR-KMGA's parameters configuration on the results of intensive tests. In the fitness computation, RMSE is applied as the metric scale to compare the candidates' simulated optical properties with the reference spectrum.

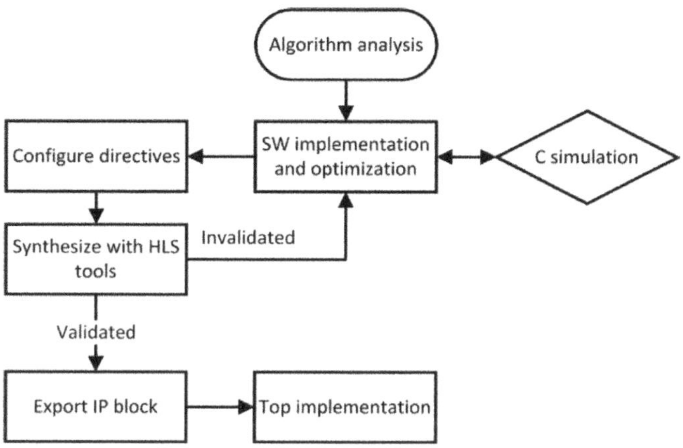

Figure 4.13. *Development process for the HCR-KMGA's FPGA implementation*

PFOA's evolution process consists of best-individual selection, prediction evolution and random evolution processes. Moreover, in order to make the routine synthesizable, the standard C function *rand()*, which is not supported by the HLS tool due to the use of static variables, is replaced by a linear congruential random number generator coded manually:

$$X_{n+1} = (aX_n + c)\ mod\ m \qquad [4.33]$$

where X is the sequence of random values, m ($m > 0$) is the modulus, a ($0 < a < m$) is the multiplier, c ($0 <= c < m$) is the increment and X_0 ($0 <= X_0 <$ m) is the start value (also called seed).

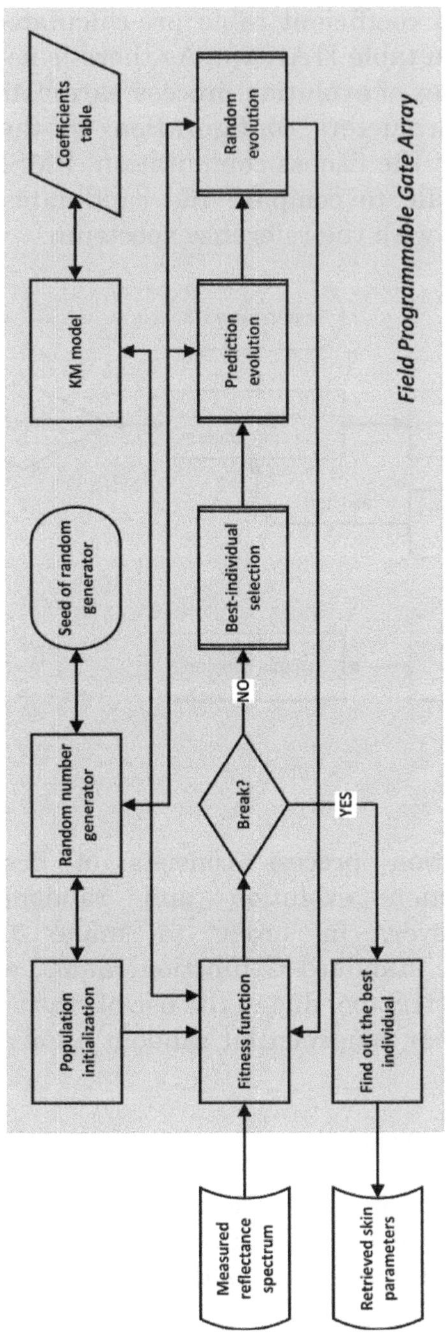

Figure 4.14. *Software implementation of HCR-KMGA algorithm for the FPGA device*

4.3.3.3. *Parameter configuration*

As there is no golden rule to define the value of the GA parameters, we perform the configuration through a series of tests. Table 4.5 compares the parameter configurations of KMGA and HCR-KMGA.

Parameters	KMGA	HCR-KMGA
Population size	100	100
Best individuals	10	1
Random selection individuals	30	-
Crossing individuals	30 pairs	-
Mutation individuals	3	-
Prediction individuals	-	49
Random individuals	-	50
Maximum fitness plateau	-	6
Minimum satisfies fitness level	-	2E-4
Max generation number	25	80
Spectrum size	34	34

Table 4.5. *Population parameter configuration for KMGA and HCR-KMGA*

Both the crossing and mutation of KMGA were selected from the literature [SYS 89] and take the probability value Pc = 0.6 and Pm = 0.02, respectively. We did not want to select a high mutation probability as it is noticed that the convergence was not smooth and often temporarily increased.

4.3.4. *Implementation evaluation experiments*

According to the algorithm described in sections 4.3.1 and 4.3.2, we can see that the KMGA and HCR-KMGA are highly parallelizable at the pixel level but the computation of each pixel is much less parallelizable and intensive due to the iteration structure of the evolutionary algorithm and complex KM function. Therefore, besides FPGA, the CPU implementation of KMGA is selected as the reference in order to obtain an unbiased evaluation. In this section, we first introduce the reference CPU implementation and then analyze the performance gains due to the algorithm improvement. Finally, the performances of the proposed FPGA design and the reference implementation are compared.

4.3.4.1. *Reference optimizations: parallel computing within GPP*

We select the GPP implementations of KMGA and HCR-KMGA as references. Both of them are realized on a Quad-Core CPU Q6600 (Intel CoreTM Quad CPU Q6600 @ 2.40 GHz X 4) with ICC 13.1.1 in 32-bit in the Ubuntu 13.04 system. In order to obtain an unbiased comparison, they are optimized using recent parallel computing technique *Pthreads* to achieve as many as possible potential performance gains from the target multi-core CPUs.

For the purpose of achieving more effective use of hardware resources in a multiple-core environment, many parallel programming methods have been developed [AYG 09, KYR 11, STR 08]. A series of studies have demonstrated that the parallelization architecture can greatly improve the image processing prototypes' efficiency [AKG 14, BAR 14, BIS 13, HOS 13, TOL 09]. *Pthreads* is one of the most popular parallel programming technologies for UNIX-like operation systems. This method allows multiplying the speed of the designs by simultaneously executing multiple threads depending on the available core number of the target processor. Comparing to the other methods, *Pthreads*' advantages consist of the rapid thread creation [BAR 17], the shared global memory and the private data zone for each thread, so the *Pthreads* implementations may gain more potential improvement in the running time and hardware cost performances.

In parallel programming, the first step is usually to find out the independent data flows in the original algorithm. Figure 4.15 illustrates that KMGA method analyses each single pixel's reflectance spectrum. That is, the algorithm sweeps all over the lesion zone pixel by pixel, and their data flows are independent of each other. This intrinsic nature of KMGA provides a nice parallelization potential.

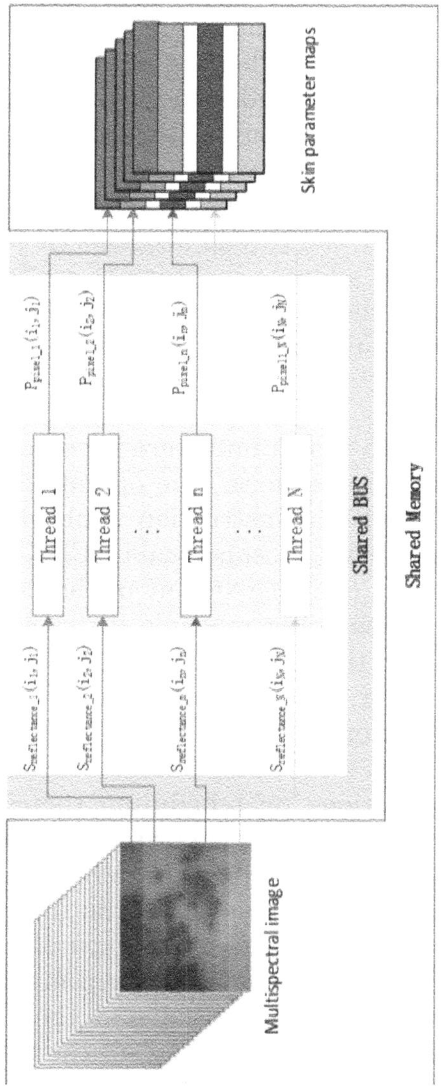

Figure 4.15. *N-threads KMGA architecture using POSIX threads: $S_{reflectance\_n}$ is the reflectance spectrum at (i_n, j_n) in the work area of the nth thread and P_{pixel_n} is its retrieved skin parameters. For a color version of the figure, see www.iste.co.uk/li/image.zip*

Using the *Pthreads* technology, we designed a single program multidata parallelization as shown in Figure 4.15. Within this architecture, the multispectral image and retrieved skin parameter maps are stored in two allocated shared memory regions in floating-point format. Meanwhile, the original image is divided into *N* work areas and each of them is distributed to a single thread. Since all the data flows are independent, there are no interactions between two threads. Once the processing of a distributed reflectance spectrum is finished, the thread accesses the shared memory again with a store address in order to write data into the right position and then starts another operation. During the processing, parallel threads need only to traverse their flown data segment.

In addition, given that a multicore superscalar processor is selected as hardware platform, the operations are scheduled according to a multiple instruction multiple data (MIMD) strategy, which allows a superscalar CPU architecture to execute more than one instruction during a clock cycle by simultaneously dispatching multiple instructions to different functional units on the processor such as an arithmetic logic unit, a bit shifter, or a multiplier. For example, the computations of equation [4.23] consist of two independent terms $f_{mel}\ U_{a.melanosome}$ and $(1\text{-}f_{mel})\ U_{a.baseline}$. MIMD can therefore accelerate computations by scheduling their instructions in parallel rather than make the algorithm run them one by one via a superscalar CPU architecture. Moreover, *Pthreads* can be automatically scheduled depending on the cores number in an out-of-order execution way by the compiler, therefore the developers have no need to consider the compatibility between the thread numbers implemented and the core number to make the processor run at full load all the time. This allows an effective use of processors resources and an important acceleration of the desired operations.

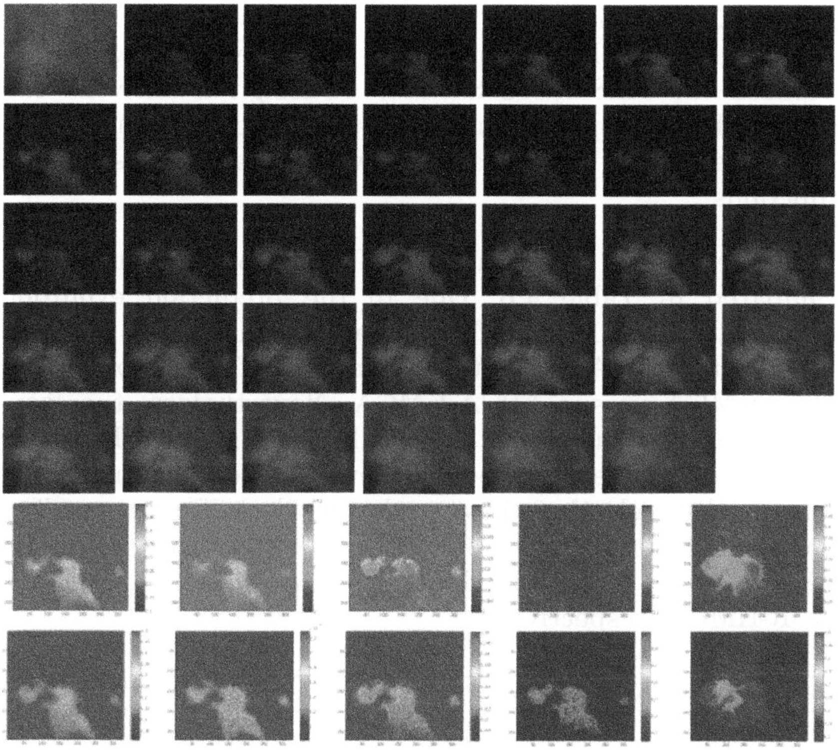

Figure 4.16. *Multispectral image measured by ASCLEPIOS from 450 to 780 nm with a step of 10 nm (top), and simulation results of five maps obtained by KMGA (middle) and by HCR-KMGA (bottom). These five maps (from left to right) consist of melanin concentration, epidermis thickness, volume blood fraction, oxygen saturation and dermis thickness. For a color version of the figure, see www.iste.co.uk/li/image.zip*

4.3.4.2. *Comparison experiments*

Figure 4.16 displays the skin parameter maps retrieved by KMGA and HCR-KMGA from a standard multi-spectral image. Obviously, the latter makes less noise from the visual effects and provides a clearer skin lesion zone for diagnostics. According to the test results, we can find that the HCR-KMGA implementations have lower fitness values than the KMGA ones (2.6×10^{-4} vs. 3.4×10^{-4}; lower is better). Meanwhile, note that the fitness values are almost identical to the FPGA and CPU implementations. This is because the HLS-based SW/HW co-design framework can effectively transplant an algorithm specified in C-like languages onto FPGAs almost without any omissions of functions.

The efficiency performances of all implementations are compared in Table 4.6. Due to the prediction evolution strategy, HCR-KMGA offers an acceleration gain of 2.28× relative to KMGA, while there is a gain of 2.13× for FPGA-HCR-KMGA versus FPGA-KMGA. Nevertheless, the loop-level and instruction-level parallelism enable FPGAs to appear to provide much better hardware performances than CPUs, although they have a lower clock frequency. The speed gains due to the platform are 5.84× and 5.45× for FPGA-KMGA versus KMGA and FPGA-HCR-KMGA versus HCR-KMGA, respectively.

KMGA	HCR-KMGA	FPGA-KMGA	FPGA-HCR KMGA
1	2.28	5.84	12.43

Table 4.6. *Implementation acceleration ratio comparison: the clock of CPU and FPGA are, respectively, 2.4 GHz and 50 MHz*

Finally, we compare the hardware resources consummation of the two FPGA implementations in Table 4.7. This comparison indicates that HCR-KMGA consumes much less RAM than KMGA. This is because the data size of the population are well reduced according to the approach presented in section 4.3.2; it therefore need not allocate as much storage space as before. In contrast, HCR-KMGA consumes more than other components; this is because PFOA has a more complex architecture than GA and HLS has to spend more resources for the operation control flow. However, this difference is very tiny; it can almost be ignored in practical applications.

Components	FPGA-KMGA	FPGA-HCR-KMGA
BRAM_18K	192	32
DSP 48E	2352	2431
Flip-Flop	467264	493177
LUT	668784	712894

Table 4.7. *Used hardware resources estimation of FPGA-KMGA and FPGA-HCR-KMGA implementations on Virtex7-XC7VX1140T of Xilinx*

4.4. Discussion

This chapter explores the fundamental concepts of HLS technique according to an example of an embedded skin lesion diagnostic application, which combines a complicated skin light propagation model, KM model, and a high running cost function optimizer genetic algorithm. Historically, this case should have required a significant effort cost to be implemented onto FPGAs even for an experienced FPGA engineer due to the large gap between the algorithm complexity and the low abstraction level of register-level languages. However, with the help of HLS, we do not have to suffer of the complexity of the low-level logic control and data flow any more. Nevertheless, since the behavior block is specified in standard C language and synthesized automatically, we can quickly switch the desired compilers (ICC and Vivado_HLS) without manually respecifying the behavior block in RTL after the function verification in the software environment. This allows us to put more attention into algorithm and architecture improvement.

Thanks to the benefits of HLS, we can focus exclusively on the desired behavior description and propose a new HCR-KMGA. A series of complicated optimizations are made on the target design. According to the algorithm evaluation experiments, the new proposed algorithm doubles the convergence speed of the evolution process and achieves a higher accuracy.

In addition, the operation scheduling of the target implementation is optimized by using HLS directives. The final RTL is compared with the multicore implementation optimized through *Pthreads*. The experiments demonstrate that FPGAs achieve a speedup of around 6× in hardware terms. However, this optimizing process is still painful for software engineers who are not familiar with HLS and FPGAs because RTLs are constrained by hardware resources and the features of the target devices that differ from the GPP. In Chapter 5, we further discuss the HLS-based design flows and propose a code and directive manipulation strategy for efficiency optimizing.

5

CDMS4HLS: A Novel Source-To-Source Compilation Strategy for HLS-Based FPGA Design

High-level synthesis (HLS) is one of the most recent synthesis techniques for FPGA-based designs. It can improve the performances of the designs, especially for real-time image/video processing developments. Nevertheless, in order to free the users from the complex coding rules and hardware constraints, an improved HLS design flow has emerged. This flow combines the source-to-source (S2S) compilers with HLS tools in order to automatically improve the efficiency of the source code. The motivation is to enable designers to concentrate their attention on the algorithm descriptions rather than hardware optimizations. Therefore, how to make compilers generate more efficient code for HLS becomes a new challenge. In this chapter, we present a novel HLS source transformation technic, Code and Directives Manipulation Strategy for HLS (CDMS4HLS), which optimizes the HLS source code in variant hierarchies by using different parallel computing strategies. We illustrate how this approach can effectively improve design performance and achieve more acceleration gains than the reference design flows.

5.1. S2S compiler-based HLS design framework

Figure 5.1(a) illustrates the framework of classical HLS-based FPGA designs. First of all, designers specify the software prototype of targeted algorithm in C-like languages and debug it in a test bench using common C compilers. Next, the confirmed code is imported into an HLS tools as original source for C-to-RTL synthesis. During this process, the designers configure the synthesis constraints/directives to make their implementations suitable for different design requirements. Finally, the generated HDL specification is simulated by an RTL simulator like ModelSim, and then exported as IP-blocks. This approach does not require specific knowledge of both software and hardware, so users can concentrate their attentions only on the algorithm specifications in high levels. However, with the widespread use of HLS tools in the embedded system world, more issues related to time control, execution speed and consummation, etc., emerge. In order to discover the best design solution, designers have to repeatedly configure the directives, and sometimes even have to respecify their algorithms to ensure the input sources are detectable by HLS processes. This is a quite painful and time-consuming job even for an experienced SW/HW codesign engineer.

The issue of classical HLS design flow discussed above is caused by three major reasons: (1) the C-language suitable for HLS tools is just a subset of C-like languages; we therefore cannot benefit from all of the C advantages during the algorithm specification; (2) different designers and algorithms may result in different code structures and styles, so the data dependency of input sources sometimes cannot be perfectly determined by scheduling process for optimization and (c) the existing synthesis tools provide dozens of directives for hardware constraints that require the designers to be quite familiar with synthesis tools.

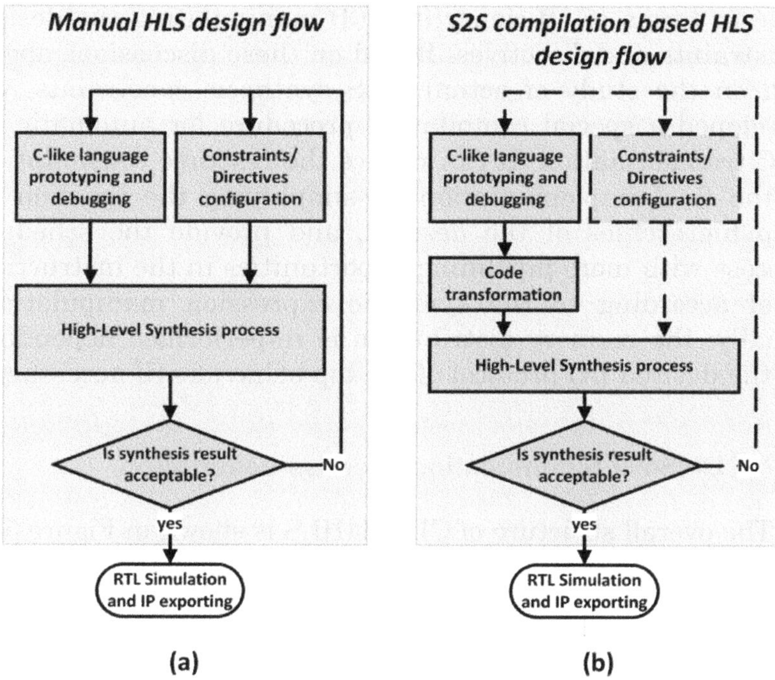

Figure 5.1. *Manual and source-to-source compiler-based HLS design framework*

For the purpose of further simplifying the development cycles, we present here a S2S transformation based HLS flow (see Figure 5.1(b)). In this new framework, a S2S compiler is inserted between the software prototyping and synthesis process. In contrast with the manual specification and configuration, this bridge tool automatically transforms the original code into the more efficient sources. For example, Alle *et al.* propose an efficiency improvement approach for loop pipelining in HLS through a semiautomatic S2S transformation in [ALL 13].

An ideal S2S compiler for HLS should have three abilities: (1) translating the prototype of the desired algorithm into the C language available for HLS, (2) optimizing the source code in terms of efficiency and (3) manipulating the design constraints and directives. Based on these discussions above and on the study of actual HLS synthesis constraints, we developed a special compilation procedure for automatic C-to-C transformation. It can reduce the resource consumption of the final implementation by simplifying the function or loop hierarchies of the designs, and provide the schedule process with more pipelining opportunities in the instruction level according to the symbolic expression manipulation. Finally, the memory distribution is re-performed depending on the desired I/O protocol of the top behaviors (if necessary).

5.2. CDMS4HLS compilation process description

The overall structure of CDMS4HLS is shown in Figure 5.2, which consists of function inline, loop manipulation, symbolic expression manipulation, loop unwinding and array transformation. These five different optimization forms are applied on the input source code in the proper order in order to enable each optimization technique to have maximum effectiveness.

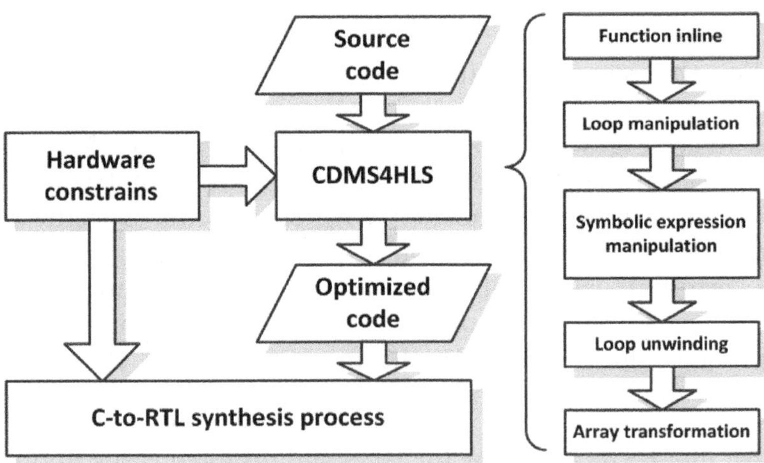

Figure 5.2. *CDMS4HLS compilation process*

5.2.1. *Function inline*

Historically, using custom function well is invariably encouraged as one of the most basic necessary abilities for programmers. However, this coding habit seriously disrupts the optimizations in the instruction level in HLS. Despite the ability to parallelize the sub functions, HLS tend to separately process each function in order to map them to the subentities, which are interconnected via assignment interfaces. The shortcoming of this approach is that all the operations in deeper level entities cannot be executed until the higher level entities finish the jobs even if some of the former's operations could be parallelized with the latter's operations.

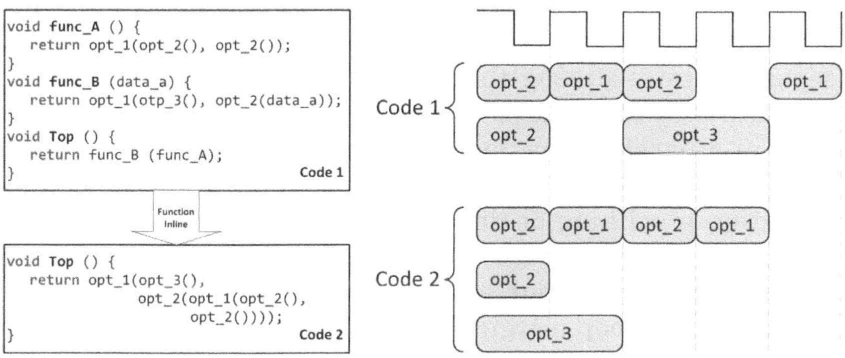

Figure 5.3. *Comparison between the sources code before and after function inline: it is assumed that opt_1 and opt_2 consume one cycle and opt_3 two cycles, and the operations of func_A and func_B are referred to by the color of red and blue, respectively. For a color version of the figure, see www.iste.co.uk/li/image.zip*

In order to offer more optimization opportunities in loop and instruction level, we first flatten the hierarchical algorithm description into a single level. Figure 5.3 shows an instance of comparing the affects from the source code before and after function inline. In *Code 1*, *func_A* and *func_B* are mapped into two separate entities, so all the operations of *func_B* are activated after the termination of *func_A*. Meanwhile, function inline assembles all the operations into a single RTL entity, therefore *opt_3* could be scheduled at the beginning of the clock sequence with other operations, which put the execution of the *opt_1* in terminal 1 cycle ahead of schedule. Thus, inlining functions can simplify the hierarchical architecture of designs and enable HLS to effect optimizations on the independent objects isolated by subfunction customizations, such as loop fusion and instruction pipeline.

5.2.2. *Loop manipulation*

Function Inline puts all the loops in the same function hierarchy. This allows a more comprehensive loop-level optimization. However, HLS processes loop-bodies as separate states during the control and datapath extraction, so successive loops have to be sequentially executed rather than in a parallel way, even if they are independent of each other. Nevertheless, separate states prevent the design from operation pipelining or data sharing. In order to potentially greater optimize the loop-body logic; CDMS4HLS manipulates the source code in loop-level using loop flattening and loop merging (see Figure 5.4).

Given that only the same-hierarchy loops can be merged together, CDMS4HLS first flattens nested loops into simple loops, respectively. Next, all the simple loops are merged into a single one. In CMDS4HLS, the largest loop bound of all the original loops is selected as the bound of the new generated loop, and their bodies are controlled and optionally executed according to *if* statements. By this means, separate loop-bodies are fused into a single state. This simplifies the hierarchical architecture of the design and offers nice opportunities to potentially greater optimize the loop-body logic. The total state transition number is defined to estimate the complexity of the FSM to be implemented. As shows in Figure 5.4, *Code 3* requires only N^2 state transitions while $2N + N^2$ for *Code* 1.

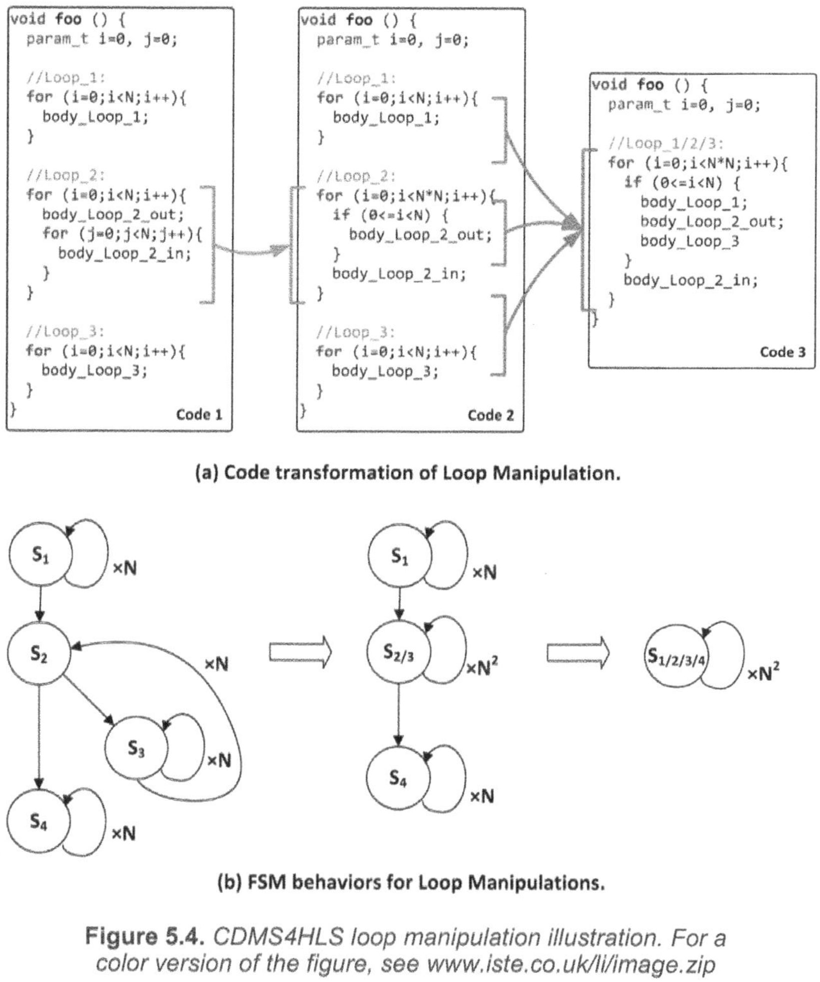

Figure 5.4. *CDMS4HLS loop manipulation illustration. For a color version of the figure, see www.iste.co.uk/li/image.zip*

5.2.3. *Symbolic expression manipulation*

The ability to analyze the source code in symbolic terms is essential for instruction-level optimizations. This can save hardware consumption and accelerate the designs by reducing operation numbers and facilitating the pipeline schedule of HLS.

Strategies	Original expressions	Equivalent expressions
Folding	$1 + 2a + 3 - 4a$	$4 - 2a$
Division	(2 x a)/(4 x b)	a/(2 x b)
Short-circuit evaluation	a && 0 && b	0
Normalization	If $(1 + a < b - 2)$	If $(a - b < - 3)$
Segmentation	Return a x b x c x d	*tmp*1 = a x b *tmp*2 = c x d return *tmp*1 x *tmp*2

Table 5.1. *Symbolic expression manipulation strategies*

Table 5.1 lists the strategies applied in the symbolic manipulation for HLS. Folding, division, short-circuit evaluation and normalization reduce a design's hardware or time consumption by simplifying computations mathematically. For example, the represented expression for folding has three additions/subtractions and two multiplications while its equivalent expression has only one subtraction and one multiplication. Nevertheless, normalization is an optimization aimed at the comparison operations. It re-performs the expression irregularly by locating the unknown variables and the constant on the different sides of the comparison operator, and then simplifies them by other strategies, respectively. Obviously, the equivalent expression consumes less hardware and cycles than the former.

Segmentation is a transformation that enhances HLS's detection ability in terms of instruction-level parallelism. For polymerization, the existing HLS tools can only pipeline different terms while scheduling the operations in a single long term in sequence. This is because a “term” is the minimum parallelizable element in HLS procedure. Thus, CDMS4HLS reperforms the long terms with numbers of multiplication/division operators (more than three operators) via segmentation to make these independent operations detectable to HLS.

5.2.4. Loop unwinding

Loop unwinding multiplies the running speed of implementations depending on the unrolling times n. The theoretical value of the acceleration ratio R should have been 2^n x. However, sometimes the efficiency improvement achieved by this optimization is lower than this expected value. This is because loop iterations always share a single top interface and the reading/writing operations of the $(i+1)$th iteration has to be delayed for certain cycles relative to the ith iteration. The value of R can be formulated as follows:

$$R(n) = \frac{L}{L_{lp}(n)}$$

where

$$L_{lp}(n) = 2^{-n} \times L + (2^{-n} - 1) \times D_{ac}$$
$$D_{ac} = \max\{D_{rd}, D_{wr}\}$$

In these equations, L and L_{lp} are the loop latencies before and after unwinding, and D_{ac}, D_{rd} and D_{wr} are the delays due to the access conflicts, reading operations and writing operations. The last equation indicates that the maximum of D_{rd} and D_{wr} is selected as D_{ac}, this is because both reading and writing produce access conflicts, but whatever the cause, loop unwinding delays the entire iteration. Thus, D_{ac} only needs to exceed the maximum value between D_{rd} and D_{wr}.

5.2.5. *Memory manipulation*

In practical applications, designers have to perform different memory I/O protocols to match to variant system architectures, but the specification of I/O protocols are constrained by memory allocations within HLS. For example, HLS performs only a single memory I/O port for each independent array even if multiple ones are allowed in the design. In order to maximize the efficiency gains with the different I/O protocol constraints, the last step of CDMS4HLS is to manipulate designs' memory allocations depending on the IP protocols and data types.

Generally, there are two array transformations that are frequently used in HLS: array partition and array reshape. The motivation of array partition is to increase the amount of read and write ports for the storage by partitioning an array into smaller subarrays or individual elements. This will result in RTL with multiple small memories or multiple registers instead of one large memory (see Figure 5.5(a)). Meanwhile, array reshaping combines array partitioning with array mapping to create a single new array with fewer elements but wider words (see Figure 5.5(b)). That is, it first splits the array into multiple subarrays and then recombines them into a new array with wider words. In order to maximise the potential improvement of the throughput of the design, CDMS4HLS set the maximum available port amount as the subarray amount in array partitioning, while the word width of the new array equals to the data bus width. Moreover, these two methods are usually combined. For a W-bits N-ports memory interface, CDMS4HLS first partitions the array into N sub-arrays and then reshapes each sub array into $[W = Wo]$ bits, where Wo is the word width of the original array.

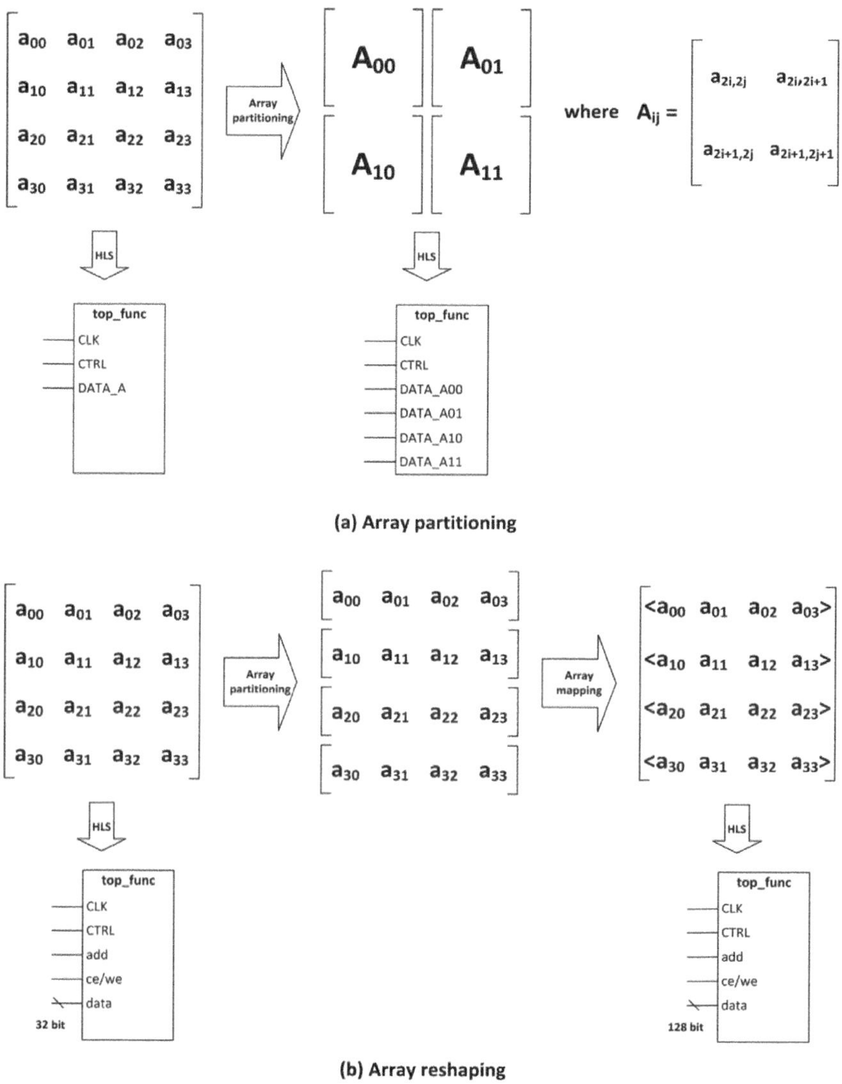

Figure 5.5. *Array transformation: [ ] refers to arrays and < > refers to the vector formed by multiple elements*

5.3. CDMS4HLS compilation process evaluation

In this section, we evaluate CDMS4HLS with four different implementations using AutoESL, which has been acquired by Xilinx and is now part of Vivado HLS. These algorithms are tested using 8 × 8 arrays with *float*, *int* and *short* data formats. A 2-ports 128-bits memory interface is set as the I/O protocol of the top behavior. In order to obtain an unbiased conclusion, the performance of these four implementations is further compared with their optimized versions using the Polyhedral Framework and Vivado HLS design suite only. All the experiments are made on the xc7a200tfbg676-2 of Xilinx.

5.3.1. *Performances improvement evaluation*

For the first step of the evaluation, four different algorithms are implemented and functionally verified by using C language, including 3 × 3 filter for RGB images, matrix product (MatPro), Image Segmentation using Sobel operator (ImSeg) and Stereo Matching using sum of squared difference (StMatch). Next, we transform the source code via CDMS4HLS. During this transformation procedure, the code generated within each phase is synthesized with AutoESL to evaluate the performances improvement related to the previous phase.

Table 5.2 describes the clock cycle and resource consumptions of the four implementations. First of all, compared to the original versions that are transformed from their C versions through HLS without any optimizations, the targeted RTL implementations are greatly improved in terms of cycle consumption within the hardware constraints of the evaluation board. This demonstrates that with the proposed approach, HLS tools can effectively use additional FPGA resources.

Implementations	Procedures	32-bit floating point numbers (float_32)					32-bit signed integer numbers (int_32)					16-bit signed integer numbers (int_16)				
		Cycles	BRAM	DSP	FF	LUT	Cycles	BRAM	DSP	FF	LUT	Cycles	BRAM	DSP	FF	LUT
3 × 3 Filter	Original	10611	0	2	1237	1241	2355	0	12	912	1446	1779	0	6	497	882
	FI	10611	0	2	1237	1241	2355	0	12	912	1446	1779	0	6	497	882
	LM	3521	0	6	2718	3284	961	0	12	1111	1382	577	0	3	376	644
	SEM	2433	0	6	3582	3284	961	0	12	1111	1382	577	0	3	376	644
	LU	74	0	98	28540	34409	71	0	384	24724	39801	44	0	192	18394	35920
	MM	67	0	120	37583	40287	249	0	384	41241	111934	36	0	192	20551	39492
MatPro	Original	5777	0	5	534	387	4241	0	4	118	108	1681	0	1	56	55
	FI	5777	0	5	534	387	4241	0	4	118	108	1681	0	1	56	55
	LM	5633	0	5	523	368	4097	0	4	107	89	1537	0	1	41	36
	SEM	5633	0	5	523	368	4097	0	4	107	89	1537	0	1	41	36
	LU	328	0	7	1379	811	15	0	8	1021	1784	11	0	5	355	450
	MM	325	0	22	4209	2778	9	0	32	1676	2120	6	0	8	867	944
ImSeg	Original	16644	3	12	4900	4786	10844	3	12	3981	4175	9604	3	3	3356	3677
	FI	15876	3	10	4267	4295	10500	3	28	4065	4771	9220	3	7	3330	3933
	LM	6273	0	30	7945	8033	3841	0	28	9127	10046	3201	0	7	8357	9208
	SEM	5633	0	42	9463	9635	3841	0	28	9127	10046	3201	0	7	8357	9208
	LU	131	0	213	66540	86889	301	0	448	35943	41117	94	0	448	72486	81162
	MM	120	0	414	121260	126995	1960	0	445	29113	33512	824	0	224	34403	39694
StMatch	Original	37713	0	15	5350	5247	20561	0	36	4065	4071					
	FI	36177	0	10	4500	4810	20561	0	36	4065	4071					
	LM	34705	0	10	4517	4813	20497	0	36	3846	3849					
	SEM	34705	0	10	4517	4813	20497	0	36	3846	3849					
	LU	325	0	197	83748	105937	277	0	516	48844	53820					
	MM	325	0	217	87841	127550	155	0	709	84789	101804					

Table 5.2. *Performances and resources consumption evaluation (FI, function inline; LM, loop manipulation; SEM, symbolic expression manipulation; LU, loop unwinding; MM, memory manipulation)*

We can see that float_32 FI versions of ImSeg achieves a 1.05× speedup relative to its original version but consumes only 90.12% of hardware resources on average, while the float_32 LM versions of MatPro consumes hardware resources of 97.7% on average for a 1.026× speedup relative to its function inline version. This is because the control architectures of the prototypes are well simplified according to function inline and loop manipulation, allowing avoidance of unnecessary cycle and resource consumption due to the control flow. Additionally, simplifying the control architecture can also greatly accelerate the design by reducing the state transition numbers and creating more parallel computation opportunities. This can be proved by the LM versions of 3 × 3 Filter and ImSeg, which, respectively, achieve 2.85× and 2.83× speedups an average over the three different data formats.

Next, according to our estimations, the speedup due to the symbol expression manipulation is 1.447× at most in all the implementations versions. This demonstrates that it can effectively improve the computing efficiency of the design. However, it is also found that this optimization is not effective in some cases. This is mainly due to two reasons: (1) the input code has been in the simplest form (i.e. the SEM versions of MaPro and StMatch) and (2) the latencies of the operation components are too tiny to effect the performance improvement (observed from comparison between the *float* and *int* SEM versions of 3 × 3 Filter and ImSeg).

Third, the implementations are further accelerated by parallelizing the loop iterations. The execution speed of all four implementations is multiplied by unwinding the loops. According to our test, the LU optimized code, respectively, achieves 19.84×, 143.34×, 29.94× and 90.39× speedups on average for the four implementations. However, it should be noted that the resource consumption is multiplied as well within LU transformations. This results in the surface of the target implementation probably exceeding the hardware constraints. For example, the efficiency of the *int* 32 LU version of 3 × 3 filter could reach 50 cycles if its loops are completely unrolled, but due to the limitation of the DSP number (768 required vs. 740 in maximum), they can be only partly parallelized in order to make the surface of the desired implementation available to the target board.

Finally, the memory of the implementations is reallocated depending on the I/O protocol in order to maximize the data throughput of the interface. Our tests indicate that this optimization may lead to two different changes to the final implementations depending on the target algorithm and data format. The first result is that the designs are accelerated, such as the *int* 32 and *int* 16 versions of MatPro, etc. This is because MM offers more operation parallelism opportunities by expanding the interface throughput and reducing the access conflicts. Instead, the second result is that the designs are not accelerated and are even decelerated, i.e. the *int* 32 versions of 3 × 3 Filter and ImSeg. This is because expending memory interface has to consume much more hardware surface. Once the resource consumption of these implementations exceed the hardware constraints, CDMS4HLS only has to reduce the loop unwinding times to reduce the surface of the target implementations. For example, the *int* 32 MM version of ImSeg could offer an acceleration gain of 3.31× relative to its LU version if its loops are completely unrolled, but its average resource consumption would be 1.57 times as much as the available resources of the target device. Therefore, its loop unwinding time is reduced by one in order to decrease its surface.

5.3.2. *Comparison experiment*

We compare here the CDMS4HLS with two other functionally similar design flows using the source code of the four chosen algorithms (see Figure 5.6). We base the first design flow on two improved conventional S2S C/C++ compilers, an improved PoCC polyhedral framework [ZUO 13, POC 13, LI 14] and the Generic Compiler Suite (GeCoS) [MOR 08, MOR 11] (defined as PolyComp), while the other one is based on the original Vivado HLS Design Suite [XIL 12b].

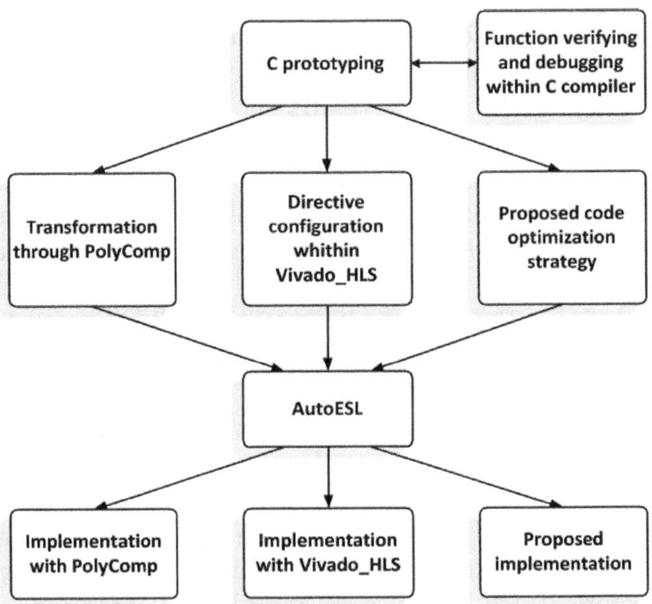

Figure 5.6. *Implementing flow with different code optimization methods, including PolyComp, manual directive configuration within Vivado_HLS and CDMS4HLS*

In order to obtain an unbiased conclusion, all the source codes are synthesized using AutoESL and their data formats are normalized to 32-bit integer numbers. Considering that PolyComp does not have the ability of I/O interface manipulation, we set the I/O protocol of the target implementations as the default of the HLS tool used.

The latency speedups of the three design flows with different algorithms are compared in Figure 5.7. This result is normalized to the *int* 32 original versions of the related algorithms. These three approaches, respectively, achieve an average of 19.01×, 22.19× and 106.54× speedups. This demonstrates that the CDMS4HLS approach can gain more performance improvements in terms of latency consumption.

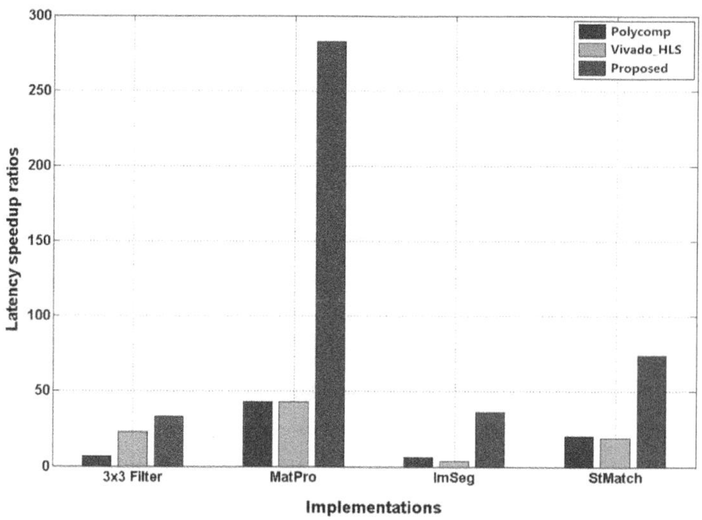

Figure 5.7. *Latency speedup comparison. For a color version of the figure, see www.iste.co.uk/li/image.zip*

Compared with the other designs, CDMS4HLS has the ability to manipulate the source code in a lower instruction level, which provides more optimization opportunities to HLS tools. Furthermore, this method can effectively reduce the transition number of the FSM behaviors of the target implementations. For example, the transition number of the MatPro optimized by CDMS4HLS is only half as many as PolyComp and Vivado HLS, respectively. In addition, it should be noted that the acceleration gains due to the interface expanding are not taken into account. That is, CDMS4HLS and Vivado HLS may achieve more speedups than Poly-Comp.

5.4. Discussion

The CDMS4HLS approach improves the performance of the desired designs in various ways, including function and loop hierarchy optimization, symbolic expression manipulation and memory/interface protocol manipulation. These methods allow a much simpler FSM architecture for the final implementations and provide more operation pipelining opportunities to the HLS tools.

Related to PolyComp, CDMS4HLS is a compilation strategy specially designed for HLS processes, therefore it can more effectively optimize the control behavior of the C prototype, which can provide more optimization opportunities in terms of instruction-level parallelism. Meanwhile, Vivado HLS does not have the ability to manipulate the symbolic expression of the source code, while CDMS4HLS can transform them into forms that can be detectable by HLS tools. The experimental results demonstrate that the latter can greatly reduce the cycle consumptions of designs and enable the HLS tools to effectively use additional FPGA resources. Compared with other conventional S2S compilers or the manual optimization procedures, CDMS4HLS achieves more acceleration gains. It is therefore concluded that CDMS4HLS is an effective code optimization strategy, which substantially improves the HLS-based FPGA designs.

In this chapter, CDMS4HLS has been illustrated and evaluated using four basic image processing algorithms. It is interesting and necessary to apply this improved FPGA design flow into some complex real-time image processing applications. Since computationally intensive algorithms may result in a complicated control flow and operation scheduling, more efforts are usually required for development and optimization. Testing and verifying CDMS4HLS according to a practical and complex case can further evaluate its feasibility for commercialization.

6

Embedded Implementation of VHR Satellite Image Segmentation

Remote sensing techniques are increasingly used in geological exploration, natural disaster prevention, monitoring, etc. They usually require very high resolution (VHR) satellite images, which are texturally rich and may raise the running cost of systems. In some cases, for example during volcanic eruptions or flood monitoring, fast and effective processing methods are necessary because the information needs to be extracted and considered as fast as possible. However, image segmentation models usually used to detect objects or other relevant features in VHR satellite images are fundamentally time consuming, which is a real hassle for researchers.

In this chapter, we present a texture region segmentation method for VHR satellite images in a high-abstraction C environment and realize its RTL FPGA implementation with the CDMS4HLS design flow described in the previous chapter. First of all, we base the design on a promising level set method (LSM) segmentation dedicated to VHR satellite images [AKB 12, HIC 13, HUA 11]. Next, the lattice Boltzmann method (LBM) is used as the solver of the level set equation for its highly parallelizable capability. Finally, the algorithm is implemented into the register-transfer level for FPGA using the improved HLS tools.

6.1. LSM description

6.1.1. *Background*

The LSM refers to the class of active contour models that use the implicit representation of the evolving curve instead of the parametric one, i.e. the Lagrangian framework and Eulerian framework [KAS 88, SAM 00]. Its advantages include the following:

– it allows the texture of interest to be selected and segmented depending on the user requirements by providing a set of user-controlled parameters;

– it allows easier handling of complex shapes and topological changes;

– it can convert the images from 2D to 3D straightforwardly;

– it allows constraints on the smoothness of the boundaries to be added easily via regularization terms.

The basic idea of the LSM is to continuously evolve the zero level of a given level set function in the image domain until it reaches the boundaries of the interested regions or objects. First of all, we define a closed curve in the image. Next, the distance between the boundary of this curve and each pixel $d(x,y)$ is expressed by using a signed distance function:

$$d(x,y) = \phi(x,y,t) \qquad [6.1]$$

where (x, y) refers to the pixel and t is time. We can see that at any time t, the pixels on the curve always satisfy:

$$\phi(x,y,t) = 0 \qquad [6.2]$$

Differentiating the equation above with respect to t, we can obtain:

$$\phi_t + \phi_x \frac{dx}{dt} + \phi_y \frac{dy}{dt}=0 \quad [6.3]$$

According to the chain rule, we can figure out the following equation:

$$\frac{\partial \phi}{\partial t} = V|\nabla\phi| \quad [6.4]$$

where ϕ, $\nabla\phi$ and V are, respectively, the level set function, the gradient of ϕ and the speed function that should drive and attract the evolution of the active contour toward the object boundaries. Equation [6.4] is known as the level set equation, which can be used to govern the active curve evolution. According to whether the speed function uses local or regional statistics, the LSMs can be divided into two general approaches: edge-based and region-based methods. The first one uses an edge indicator depending on the gradient of the image as in the classical snakes[1], but these methods are effective only when the boundaries of the object have a clear change in gray value. Moreover, they are sensitive to noise and ineffective when the object of interest is without edges. The second one uses some regional attributes to stop the evolving curve. These methods are robust against noise and effective even if the object is without edge.

Many efforts have been made to apply the geometric active contour frameworks to the field of remote sensing. For example, Karantzalos *et al.* [KAR 09] propose a variational geometric level set functional for man-made object detection from aerial and satellite images, while Ball *et al.* [BAL 07] present a supervised hyperspectral classification procedure consisting of an initial distance-based segmentation method

1 A snake corresponds to an energy minimizing spline guided by external constraint forces and pulled by image forces toward features: edge detection, subjective contours, etc. Basically, snakes are trying to match a deformable model to an image by means of energy minimization.

that uses best band analysis, followed by a level set enhancement that forces localized region homogeneity. However, none of these methods are local and, therefore, they cannot benefit from high-performance computing techniques. The main limitation is due to the fact that, at each iteration, the average intensities inside and outside the contour have to be calculated, thus dramatically increasing the processing time by increasing communications between processors.

Recently, some researchers pointed out that the LBM is a potential approach to solve the level set equations [CHE 07]. This method was originally designed to simulate Navier–Stokes equations for an incompressible fluid [HE 97] and was only recently used in image segmentation [BAL 15]. Differing from the other solvers [BAL 07], the LBM exhibits local and not global computations, which may result in less data dependencies than the other methods. Given the processing devices available today, such as FPGAs and GPUs, which feature powerful parallel programming, this method appears as a natural candidate to improve the segmentation performance because of numerous advantages, such as simplicity, its intrinsic massively parallel nature and second-order accuracy both in time and space.

6.1.2. *Level set equation*

We perform the following speed function by combining LSM with multikernel theory:

$$V(x) = \lambda(\varepsilon - \left(T\left(\varphi(x)\right) - T(I_t)\right)^2) + V_{reg}(x) \qquad [6.5]$$

with:

$$\varphi(x) = \left(I(x), I_f(x), s(x)\right) \in R^3 \qquad [6.6]$$

where x and I_t are, respectively, the spatial variable and the intensity of a given pixel in a region of interest, λ is a user-controlled positive parameter used to accelerate the convergence of the method toward the steady state and ε is the threshold parameter to stop the evolution. V_{reg} is a regularization term, $I(x) \in R$ is the intensity of the pixel x, the two-tuple $\left(I_f(x), s(x)\right) \in R^2$ is a simple descriptor of the texture information at the pixel x, in which $I_f(x)$ is the filtered intensity of the pixel x and $s(x)$ is the standard variance of the intensities of the neighbors of x, and T is a transformation function defined as follows:

$$K(x,y) =< T(x), T(y) > \qquad [6.7]$$

where K is a non-negative combination of two Mercer kernels k_1 and k_2 :

$$K = k_1\,(x,y) + \alpha k_2\,(x,y) \text{ with } \alpha > 0 \qquad [6.8]$$

The most commonly used kernels include linear, polynomial and Gaussian kernels. Here, we use the Gaussian kernel and define k_1 as the pixel intensity kernel, while k_2 is the texture information kernel. They are expressed as follows:

$$k_1\,(x,y) = \exp\{-\|I(x) - I(y)\|^2/\sigma^2\} \qquad [6.9]$$

$$k_2\,(x,y) = \exp\{-\| \left(I_f(x), s(x)\right) - \left(I_f(y), s(y)\right) \|^2/\sigma^2\}$$

where σ is the Gaussian root mean square width of the kernel.

On the other hand, the regularization term $V_{reg}(x)$ is defined as:

$$V_{reg}(x) = \beta\big(1 - \mu(x)\big) + v div(\nabla\phi/|\nabla\phi) \qquad [6.10]$$

where div is the divergence operator, β controls the impact of the texture information on the segmentation results, v is a positive constant and μ is a fuzzy membership value that helps with deciding if the current pixel x is a boundary or not. μ is defined as follows:

$$\mu(x) = \begin{cases} 1 - \frac{\mu-\delta}{\mu}, & \mu \geq \delta \\ 1, & else \end{cases} \quad [6.11]$$

with

$$\delta = |I(x) - \bar{I}(x)| \quad [6.12]$$

where $\bar{I}$ is the mean intensity of the local pixels Ω defined as:

$$\bar{I}(x) = \frac{\int s(x,y)I(y)dy_\Omega}{\int \Omega(x,y)dy_\Omega} \quad [6.13]$$

with $s(x,y) = \begin{cases} 1, & |x-y| < r \\ 0, & else \end{cases}$ [6.14]

where r is the radius constant, μ is a positive small parameter defined as 0.1 and Ω is the image domain. Consequently, the first term of V_{reg} in equation [6.10] drives the active contour as close as possible to those pixels with which $\mu(x) = 1$, and the second term is the constraint on its length.

Since the proposed speed function tends to zero when $\left(T(\varphi(x)) - T(I_t)\right)^2$ tends to ε, which is a small value, the evolving contour will stop at a certain pixel whose intensity and texture information are equal to those of the selected region. Now, we can figure out the desired multikernel level set equation:

$$\frac{\partial \phi}{\partial t} = \lambda\left(\varepsilon - 2\left(1 + \alpha - K(\varphi(x), I_t)\right)\right) + \beta\left(1 - \mu(x)\right) + vdiv(\nabla\phi) \quad [6.15]$$

6.1.3. *LBM solver*

The D2Q5 (two special dimensions and five discrete velocity directions) LBM lattice structure shown in Figure 6.1 is used for the solver of the proposed level set equation. The level set evolution equation can be expressed as follows:

$$f_i(r + e_i,\ t + 1) = f_i(r,\ t) + \Omega_{BGK} + \frac{D}{bc^2}\,\boldsymbol{F}\,e_i \qquad [6.16]$$

where e_i and f_i are the velocity vectors at each link of the model shown in Figure 6.1, the particle distribution that moves along the corresponding link, r is the position of the cell and t is time. $\boldsymbol{F}$ is an external force, $D = 2$ is the grid dimension, $b = 5$ is the link at each grid point, $c = 1$ is the link length and Ω_{BGK} is the Bhatnagar–Gross–Krook model expressed as:

$$\Omega_{BGK} = \frac{1}{\tau}(\bar{f}_i(\vec{r}, t) - f_i(r, t)) \qquad [6.17]$$

where τ is the relaxation time and $\bar{f}_i$ is the local Maxwell Bolzmann equilibrium particle distribution function which depends on the particle and macroscopic velocity, macroscopic fluid density, and on the thermodynamic temperature. In the work of Zhao [ZHA 15], this equilibrium distribution is considered as a typical diffusion phenomenon. According to the Chapman–Enskog analysis, the following diffusion equation can be obtained from the LBM evolution:

$$\frac{\partial\rho}{\partial t} = \xi div(\nabla\rho) + F \qquad [6.18]$$

where ξ is the diffusion coefficient. It is obvious that equation [6.18] is similar to equation [6.15] with the body force expression as follows:

$$F = \lambda\left(\varepsilon - 2\big(1 + \alpha - K(\varphi(x), I_t)\big)\right) + \beta\big(1 - \mu(x)\big) \qquad [6.19]$$

Now, we can obtain the desired level set equation solver:

$$f_i(r + e_i, t + 1) = f_i(r, t) + \frac{1}{\tau}\left(\bar{f}_i(\vec{r}, t) - f_i(r, t)\right) + \frac{D}{bc^2}\left(\lambda\left(\varepsilon - 2\left(1 + \alpha - K(\varphi(x), I_t)\right)\right) + \beta\left(1 - \mu(x)\right)\right) \quad [6.20]$$

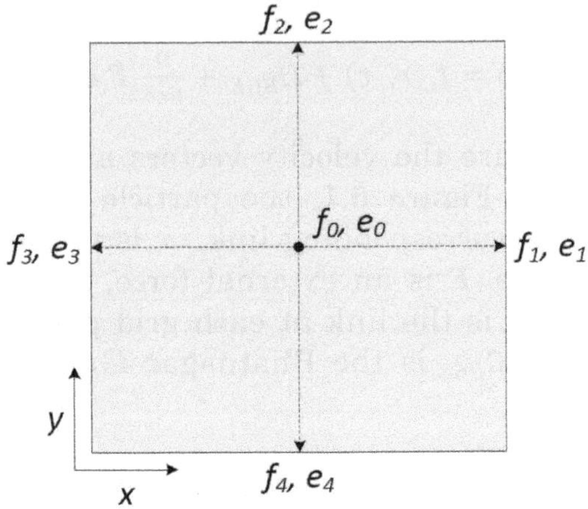

Figure 6.1. *The D2Q5 lattice Boltzmann method (LBM) model*

6.2. Implementation and optimization presentation

6.2.1. *Design flow description*

Figure 6.2 shows the design flow used to implement the LBM onto an FPGA device. Considering most designs begin with the algorithm analysis, which requires a very high level environment to facilitate the descriptions of mathematical operations and the simulations of the model, we select MATLAB as the favored user-level design environment for its powerful built-in image processing toolbox and advantages in terms of vector processing.

Once the algorithm is verified, we can start to prototype the algorithm for FPGA synthesis. Because of the HLS technique, the algorithm behavior can be described in the C environment and verified by using the common C compilers.

It should be noted that the original prototype has to be debugged depending on the input code constraints of high-level synthesis due to the special architecture of FPGAs compared to general-purposed processors.

Before running the synthesis, we make some source code optimizations using the CDMS4HLS strategy proposed in the previous chapter in order to improve the design performances. Finally, the low-level code generated by the C-to-RTL conversion process is evaluated by using a hardware description language estimator to determine whether it satisfies the performance requirements. If satisfied, the generated implementation is then exported as an IP core to the test bench for top level simulation; otherwise, the process of code optimization is iterated again.

This development design flow uses five different development tools within Ubuntu, including MATLAB, Gedit, ICC, AutoELS and System Generator, which can benefit both the development productivity and design performance according to their different development environments. For example, MATLAB and System Generator are, respectively, used for algorithm analysis and top level simulation of the generated kernel for their advantages of high-level abstraction. Meanwhile, Gedit is selected as the code editor for prototyping and optimizing in a C environment, while ICC for simulating the source code by using general-purpose processors. The gap between C and register-level languages is bridged via AutoELS, which is one of the leading high-level synthesis tools for its abilities to produce high-quality register-level languages.

Within this design flow, all of the manual, necessary cycles, including algorithm analysis, prototyping, debug, optimizing and top level simulation, are made in a high-abstraction environment, while the C-to-RTL conversion is automated by incorporating the HLS process. This is an effective approach to free the users from the boring work of hardware configuration required at the low abstraction level. Moreover, this approach allows for design space exploration in different environments (MATLAB, C and System Generator) and processors (CPU, FPGA and GPU), which provides a large number of opportunities to find potential design solutions compared to design flows only targeted to a single type of environment or device.

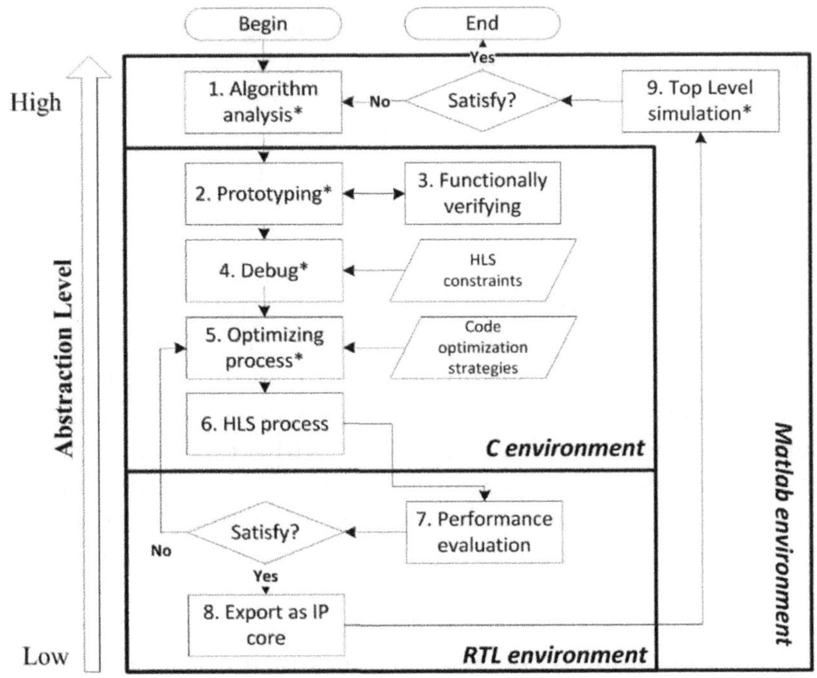

Figure 6.2. *HLS-based design flow: "*" refers to the necessary manual cycles*

6.2.2. *Algorithm analysis*

Algorithm 6.1 shows the pseudocode of the target LBM. The implementation using this original description will result in an inefficient block hierarchy and a complex control behavior, which seriously constrains the potential optimization gains of the design in terms of instruction parallelization. According to our analysis, the primary bottleneck of the optimization occurs as the loops in the routine cannot be completely unrolled due to the hardware constraints. This results iccurs asn an inefficient parallelization because the instruction-level optimizations are disrupted by the loop and function hierarchy within the HLS process. For example, the loops of Lines 2, 6 and 10 cannot be parallelized, even if they are independent, because the generated register-transfer level implementation is based on the finite state machine (FSM) architecture, and the loop bodies are processed as separate states that have to be run one by one.

Algorithm 6.1. Pseudocode of the original LBM
Require: original image I, initial zero level contour ϕ Ensure: the final zero level contour ϕ 1: Initialize the coefficients 2: for all pixels do 3: Compute Gaussian memberships with equation [6.11] 4: end for 5: Initialize the local textural information 6: for all pixels do 7: Compute the body force with equation [6.19] 8: end for 9: $\tau \leftarrow \frac{9.0\times\gamma+2.0}{4}$ with $\boldsymbol{\gamma} = \mathbf{1}$ 10: for all pixels do 11: Initialize the LBM distribution function 12: end for 13: for all iterations do 14: for all pixels do 15: Perform streaming-collisions with the D2Q5 LBM Lattice structure 16: end for 17: for all pixels do 18: Update the distance value 19: end for 20: end for

In the following sections, we will sequentially apply four different source code optimization techniques regrouped in the CDMS4HLS strategy onto the target implementation, including function inline (FI), loop manipulation (LM), symbol expression manipulation (SEM) and loop unwinding (LU).

6.2.3. *Function inline*

According to the design flow proposed in section 6.2.1, the C prototype of the design needs to be debugged for HLS. Since FPGAs structurally differ from general-purpose processors, neither the static definition nor the dynamic loop boundaries are supported in HLS-available C programming. Thus, one of the primary tasks of the debugging process is to correctly manage both the static variables and the dynamic loop boundaries in the original code. These constraints even result in some commonly used C library functions, i.e. *exp()*, *rand()* or *pow()*, not being compliant with HLS. To resolve this issue, we pragmatically build a novel C library that includes a series of alternatives for each of the C functions that are not compatible with HLS. In this work, the *exp hls()* and *pow* hls()* functions are used to, respectively, substitute the standard *exp()* and *pow()* functions, in which "*" indicates "1, 2, 3 … 10".

However, the approach presented above seriously raises the complexity of the generated RTL. In HLS, the C subfunctions are processed as reusable subblocks. Consequently, the subfunctions are separately processed first and then interconnected via assignment interfaces in the functions in the upper level. Hereby, we use the number of assignments to estimate the complexity of the implementations. Figures 6.3(a) and (b) display the block hierarchies before and after debugging. In this figure, an assignment is represented by a line arrow (e.g. from *pow()* to *ImgSeg Ori()*), so we can find that the original version leads to only six assignments and increases to 15 for the debugged version. Despite an optimization possibility by using function-level parallelism, due to the isolation of each subblock, the optimizations available in the deeper loop and instruction levels are confined to the scope of the blocks even if some different block loops or operations are able to be more efficiently manipulated. Therefore, the first step of the proposed optimization process is to compress the hierarchical architecture of the generated RTL into a single level by substituting the corresponding source code for the subfunction calls in the top function, which is known as FI. This processing can offer more optimization opportunities by merging the separate loops and operations into a single scheduling scope. Furthermore, for this reason, it is first made during the optimization process.

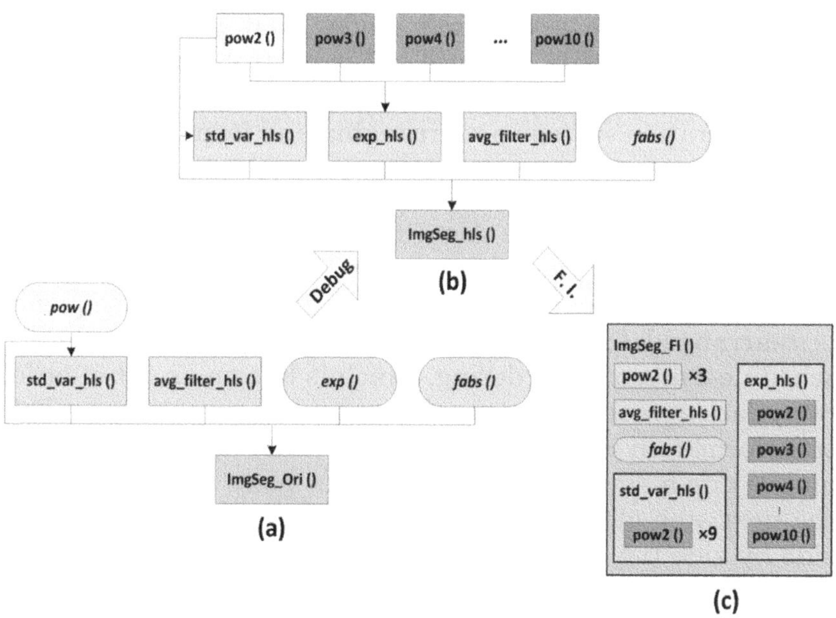

Figure 6.3. *Block hierarchy comparison: a) original version, b) debugged version and c) function inline version. For a color version of the figure, see www.iste.co.uk/li/image.zip*

6.2.4. *Loop manipulation*

In HLS, the control logic of the source code is first analyzed and extracted as an FSM. Next, the operations are assigned to the corresponding states of the control behavior to perform the control and datapath behavior. During this process, either the sequences of operations or the loop bodies are represented as a single state, i.e. Lines 5 or 9 for the operation sequence and Lines 3, 7 or 11 for the loop body in Algorithm 6.1, and the operation scheduling is confined by the state scope.

Generally, applying LM is motivated by two reasons: (1) to reduce the running-cost due to the state transit controls by reducing the state numbers and (2) to help HLS to discover more instruction-level parallelism opportunities by expanding the scheduling scopes that are confined by states. During this process, the control and datapath behavior FSM of the design is optimized by subsequently using operation sequence rearranging and loop merging.

The pseudocode of the LM-optimized implementation of this design is shown in Algorithm 6.2. Figure 6.4 displays the transition of the generated RTL within this process. After an analysis of data dependency, the operation sequence is rearranged from the math-convenient ordering into the schedule-convenient ordering, which is known as out-of-order compilation in high-performance computing. This transformation merges the different state operations (L1, 5 and 9) into a single state and provides more opportunities to the next step of loop merging by locating the separate loops side by side (i.e. L3, 7 and 11). Next, the boundary-similar neighboring loops are merged together.

According to the comparison between Figures 6.4(a) and 6.4(c), the LM process reduces the state number by 62.5% (eight vs. three transits). Meanwhile, the state transit number ratio R_{LM} can be expressed as follows:

$$R_{LM} = \frac{2+(1+ITS)\times W\times L}{(6+ITS)+(3+2\times ITS)\times W\times L} \quad [6.21]$$

where $W \times L = 948 \times 450$ is the image dimension and ITS is the maximum iteration number defined as five. Given $W \times L >> 6 + ITS > 2$, we can evaluate $R_{LM} \approx 46\%$, and the transit number of the design is therefore reduced by approximately 54%.

Algorithm 6.2. Pseudocode of the LM-optimized implementation
Require: original image I, initial zero level contour ϕ Ensure: the final zero level contour ϕ 1: Initialize the coefficients 2: Initialize the local textural information 3: $\boldsymbol{\tau} \leftarrow \frac{9.0\times\gamma+2.0}{4}$ with $\boldsymbol{\gamma} = \mathbf{1}$ 4: for all pixels do 5: Compute Gaussian memberships with equation [6.11] 6: Compute the body force with equation [6.19] 7: Initialize the LBM distribution function 8: end for 9: for all iterations do 10: for all pixels do 11: Perform streaming-collisions with the D2Q5 LBM Lattice structure 12: Update the distance value 13: end for 14: end for

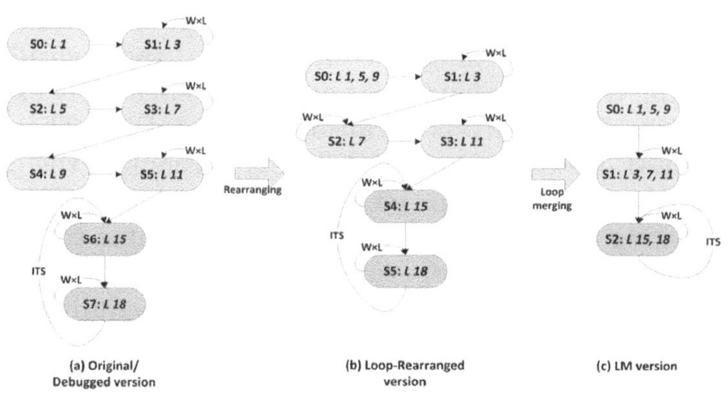

Figure 6.4. *Finite state machine (FSM) transition within loop manipulation: "L*" refers to the operations covered in the present state, in which "*" indicates the line number in Algorithm 6.1, W × L = 948 × 450 is the image dimension and ITS is the maximum iteration number defined as five*

6.2.5. *Symbol expression manipulation*

Because of the improvements of scheduling algorithms, current HLS tools allow us to optimize operation scheduling with various time or resource constraints. During this process, the parallelism in the design is first exposed according to the data flow graph and then the scheduling algorithm is used to determine the cycle within which the operations can be scheduled.

In the original version of our design, the target algorithm is mathematically described, which leads to a series of long mathematical expressions, including the computations of $Exp(x)$ and x^n, local mean intensity value $\bar{I}$ and standard variance s, shown as follows:

$$Exp = \sum_{n=0}^{\infty} \frac{x^n}{n!}$$
$$x^n = \prod_n x$$
$$\bar{I} = \frac{1}{N}\sum_{i=1}^{N} I_i$$
$$s^2 = \frac{1}{N}\sum_{i=1}^{N}(I_i - \bar{I})^2 \qquad [6.22]$$

Such mathematical functions seriously prevent designs from benefiting from the instruction-level parallelism optimizations. According to our tests, the parallelism underlying the single equations cannot be effectively exposed by the data flow analysis performed by AutoELS. Therefore, in the SEM, we transform the long expressions in the design into short ones to help HLS to discover more potential parallelizing opportunities. Figure 6.5 illustrates the code of the function *pow4_hls()* before and after the symbol expression transformation, as well as its scheduling graph. Obviously, the segmented expressions lead to a more efficient implementation. We, respectively, apply this approach to the descriptions of equation [6.22] and achieve speedups from 24.3% up to 60%.

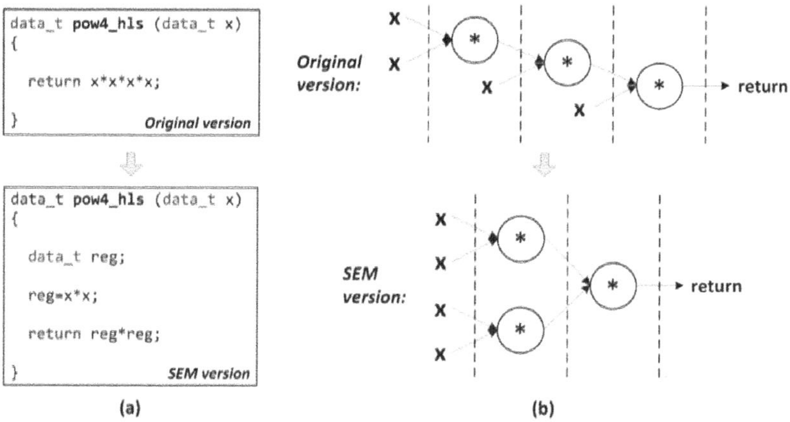

Figure 6.5. *Scheduling of pow4_hls: a) original-to-symbol expression manipulation (SEM) code transformation; b) scheduling comparison*

6.2.6. *Loop unwinding*

LU is a loop-level parallelism (LLP) form of optimization widely used in parallel computing. It first unwinds the loops in the source code and then pipelines the body operations for acceleration if their iterations are independent. By using this transformation, engineers can multiply the running speed of the design up to several hundred times. Meanwhile, it should be noted that this method is seriously constrained by the hardware resources of the target device. The area of LLP optimized version A_{llp} can be expressed as follows:

$$A_{llp} = \sum_{k=1}^{K}(DOP_k(n) \times a_k) + A_{control} \quad [6.23]$$

where K is the number of operation types, $DOP_k(n)$ is the degree of parallelism of the kth operation with the unrolling times of n, a_k is the area of the kth component and $A_{control}$ is the area of the control circuit. According to equation [6.23], the area constraint of LU can be formulated as:

$$A_{llp} < A \rightarrow \sum_{k=1}^{K}(DOP_k(n) \times a_k) < A - A_{control} \quad [6.24]$$

where A is the available area of the target device.

Besides pipelining the operations, LU also reduces access conflicts by enabling the fused iterations to share registers. Since the loop bodies are extracted as independent states in RTL, each of them has to access the memory individually even if some of the data can be shared by multiple iterations. Fusing separate iterations into a single one breaks the isolation between the loop iterations, which provides a nice opportunity to reduce access conflicts. However, it should be noted that the efficiency improvement from registers sharing among iterations is effective only when the reading operations lead to more delays than the writing operations; otherwise, the accelerations achieved by reducing reading operations will be completely offset by the delays due to writing access conflicts.

Algorithm 6.3. Pseudocode of the loop unwinding optimized implementation

```
Require: original image I, initial zero level contour ϕ
Ensure: the final zero level contour ϕ
Initialize the coefficients
Initialize the local textural information
τ ← (9.0×γ+2.0)/4 with γ = 1
for all pixels do
Compute Gaussian memberships with equation [6.11]
Compute the body force with equation [6.19]
Initialize the LBM distribution function
end for
for all iterations do
for all pixels do
#program AP unroll factor=16
#program AP pipeline
Perform streaming-collisions with the D2Q5 LBM Lattice structure
Update the distance value
end for
end for
```

Algorithm 6.3 shows the pseudocode of the LU-optimized implementation of this design. Due to the resource constraints of the target device, we only partly optimize the final nest loop in Line 10 with $n = 4$ by using the loop unroll and pipeline directives in AutoELS. Our tests demonstrate that this optimization reduces the latency of the target loop by 90.5%.

6.3. Experiment evaluation

In order to obtain an unbiased conclusion, we compare the proposed implementation with its original RTL implementation and one CPU implementation. These reference implementations are developed from the same pseudocode shown in Algorithm 6.1. For the sake of fairness, all of the reference implementations are specified within the high-abstraction environment C/C++. We name the C/C++ version *cpu_c*, *fpga_hls_ori* is the RTL generated from the source code without any optimization and *fpga_hls_opt* is the optimized implementation. Since the algorithm is complex, a large FPGA platform, Kintex 7, is selected for evaluation.

In this section, the user-controlled parameters in the algorithm are defined first. Next, its implementations are functionally verified by using four VHR satellite images taken by IKONOS or GeoEye-1. Third, the effects of the optimization technique are analyzed one by one. Finally, they are evaluated through the running-cost performance comparison.

6.3.1. *Parameter configuration*

In equation [6.20], since a D2Q5 model is used in the design, we have D = 2 and b = 5. The length of each link in the active contour model is defined as 1, so $c = 1$. The parameter λ can be used to accelerate the convergence toward the steady state, and users have to manually configure it for different input images. τ is the relaxation coefficient that is used to control the curvature, and we fix it to 2.75. α and β control the impact of the texture information on the segmentation results. In order to segment the text region of interest well, we find the suitable values of α according to observations of the segmentation results in a test image [BAL 14].

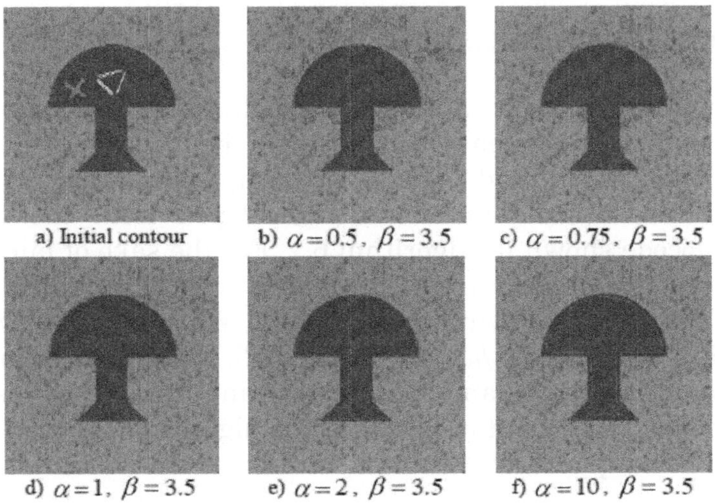

Figure 6.6. *Impact of the parameter α on the accuracy of the segmentation results (see [BAL 14]). For a color version of the figure, see www.iste.co.uk/li/image.zip*

In the result analysis of Figure 6.6, we can see that a value of α being too low (α = 0.5) leads to an oversegmentation, while a value that is too high (α = 10) decreases the precision of the result (undersegmentation). In Figure 6.7, we can also see that a too low value of β leads to an undersegmented result as the curve fails to detect the right contour. Therefore, we fixed α = 2 and β = 3.5. However, it should be noted that this configuration is obtained within double floating point numbers; for the single floating point and fixed point numbers, we changed α from 2 to 5 and 9, respectively.

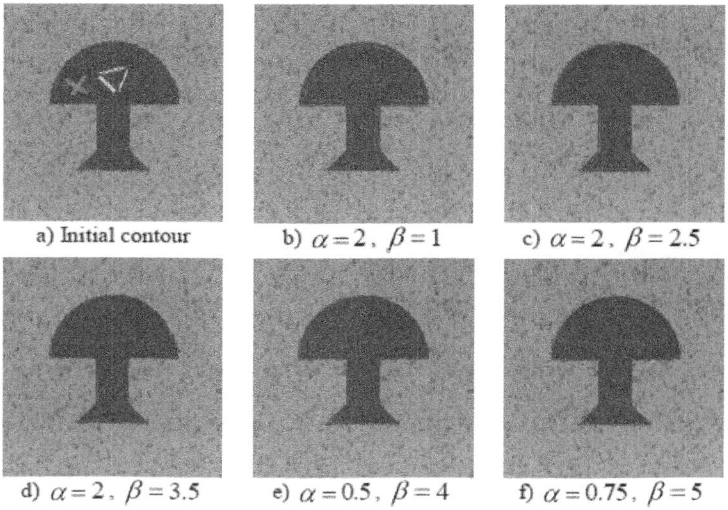

Figure 6.7. *Impact of the parameter β on the accuracy of the segmentation results (see [BAL 14]). For a color version of the figure, see www.iste.co.uk/li/image.zip*

6.3.2. *Function verification*

Figures 6.8 and 6.9 show the original images and segmentation result. Figure 6.8(a) is the image of Uxmal in Mexico taken by the IKONOS satellite in 2002. Its segmentation results demonstrate that our algorithm can effectively delimit the non-forested areas and extract the road underling the forest. Figure 6.8(b) is the image of the volcano in Iceland taken by IKONOS, as well. We can see that the activity of the volcano is well detected. The two photos of Figure 6.9 were taken by the GeoEye-1 satellite, and the ice-covered areas and the beach line are set as the intended object for the segmentations, respectively. The results demonstrate that the desired areas are well delimited despite the disturbances of the unwanted areas, i.e. the convex ices in Figure 6.9(a) or the shoreline constructions in Figure 6.9(b).

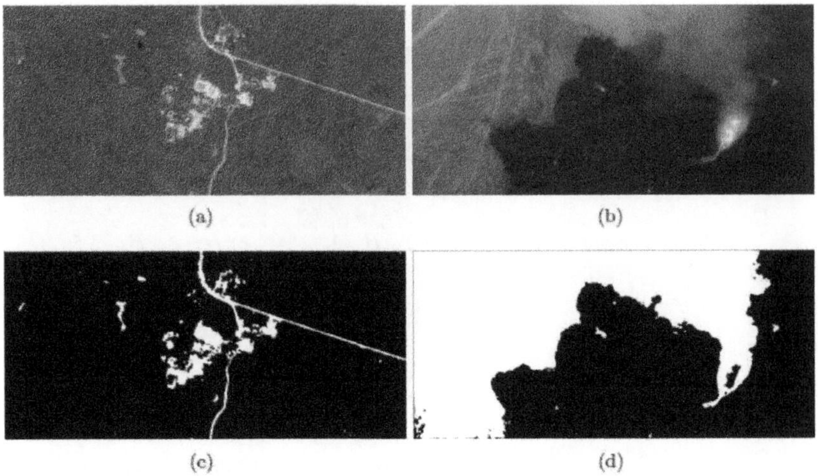

Figure 6.8. *Original images and segmentation results: taken by the IKONOS satellite: a) Original image of Uxmal; b) original image of volcano; c) segmentation result of Uxmal and d) segmentation result of volcano. For a color version of the figure, see www.iste.co.uk/li/image.zip*

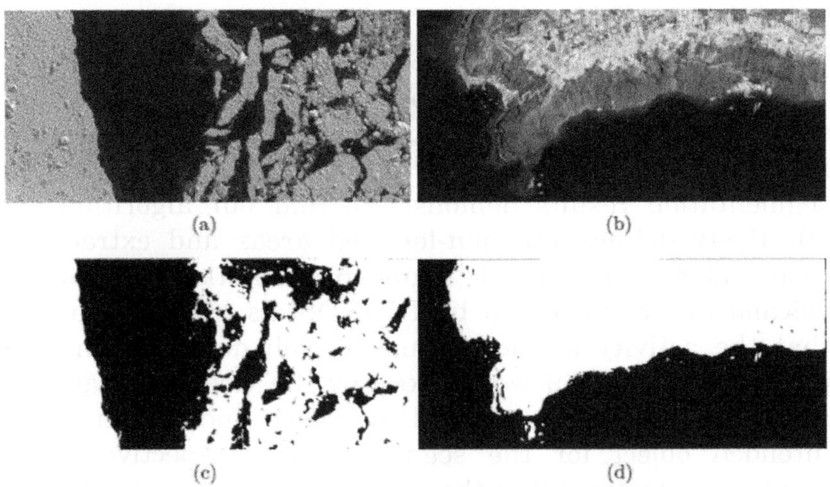

Figure 6.9. *Original images and segmentation results: taken by the GeoEye-1 satellite: a) Original image of ice sheet; b) original image of Santorin; c) segmentation result of ice sheet and d) segmentation result of Santorin. For a color version of the figure, see www.iste.co.uk/li/image.zip*

On the other hand, it is noted that the segmentation results of the different implementations are highly similar. These similarities are quantitatively evaluated by using the *corr2()* function within MATLAB and results very close to 1 have been obtained. This demonstrates that the selected HLS tool is effective to automate the C-to-RTL transplantation with an acceptable difference, and the optimizations made on the *fpga_hls_ori* would not affect the functions of the design.

6.3.3. *Optimization evaluation*

Because of the proposed design flow, we can easily discover and evaluate different design alternatives. Hereby, we subsequently apply different optimization technique onto the design depending on the nature of the HLS process. Table 6.1 details the running-cost estimation of each optimization cycle.

In Table 6.1, it is first seen that each optimization made on the design can effectively reduce the latency of the generated RTL. Their latency improvements related to the previous optimization cycle are, respectively, 14.5%, 5.1%, 8.7% and 92.7%, while the total is 94.6%. We can find that the main contributor of the overall optimization process is LU. This is because the LBM solver has a low iteration dependency, which enables the potentially high parallelism to be exploited. Therefore, the LU can multiply the running time with few data dependency constraints.

Meanwhile, it is seen as well that the hardware resource consumption varies greatly depending on the different optimization forms. FI and LM accelerate the design, as well as reduce the resource consumption. This is because they can create more operator sharing opportunities by expanding the scheduling scope confined by the function and loop hierarchy, respectively, and reduce the architectural complexity of the design, as well. In contrast, SEM and LU do not have the same capability; therefore, the generated RTLs optimized by them lead to more consumed resources. In this design, the target device is *xc7k70tfbg484-1* of *Kintex-7* from *Xilinx*, and due to the LUTs constraint, we can only partly unwind the loop. The LUTs required for each optimized implementation are around 55%, 40%, 38%, 50% and 81%, respectively.

Optimization	Latency (cycle)	Clock (ns)	BRAMs	DSPs	FFs	LUTs
Original	532,072,992	8.68	3	74	13,462	23,403
FI	454,940,736	8.58	3	43	10,607	16,916
LM	431,875,392	8.33	3	41	9858	16,161
SEM	394,141,984	8.33	3	61	12,287	21,263
LU	28,772,758	16	0	66	21,422	34,457

Table 6.1. *Optimization evaluation: original corresponds to fpga_hls_ori implementation while LU corresponds to fpga_hls_opt implementation*

Finally, it should be noted that the design is constrained by the system frequency, because highly parallelized implementations may result in long clock periods in FPGAs. The clock column of Table 6.1 shows the minimum clock periods for each implementation estimated by using AutoELS. We can see that the LU-optimized implementation requires a much longer period than the others.

Figure 6.10 compares the running-time acceleration ratio to the one of latency with a logarithmic scale. The original implementation is set as the reference in this comparison. We can see that although the efficiency performance of the design is negatively influenced by the frequency constraint, because of the significant performance gain of LU, the proposed design achieves a speedup of around 10× compared to its original version.

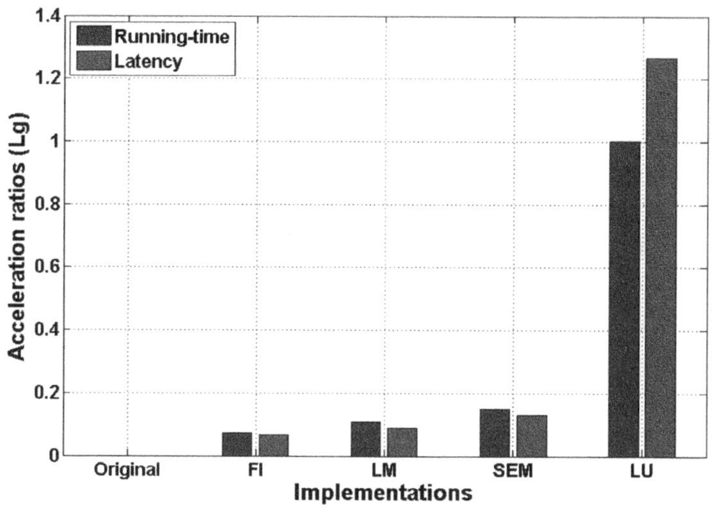

Figure 6.10. *Running time and latency acceleration (expressed in LOG) improvement of different optimized implementations. For a color version of the figure, see www.iste.co.uk/li/image.zip*

6.3.4. *Performance comparison*

This experiment estimates the overall running time of the proposed and reference implementations in the real world. The scope of this measurement covers the necessary computations and memory access operations. For CPU implementation, specific storage space is allocated in the memory for the input and output images, whereas the external RAM is used for the implementations of FPGA.

The results of this experiment are compared in Table 6.2. First of all, it is observed that *fpga_hls_ori* is 1.9× slower than *cpu_c*. The generated RTL is substantially an FSM constructed using the data flow graph of the source code, so its operation scheduling satisfies the corresponding time constraints. This prevents the design from parallel computing even if some operations are independent and parallelizable. In the experiment of section 6.3.3, it is mentioned that this version consumes only 55% of the LUTs of the target device. This demonstrates that it does not make effective use of the additional hardware resource for performance optimization. On the other hand, *cpu_c* is well optimized during the compilation, and the frequency of the target processor Q6600 is around 21× as high as the selected FPGA (2.4 GHz vs. 115.2 MHz). All of the reasons mentioned above result in the performance difference of *cpu_c* versus *fpga_hls_ori*.

cpu_c	*fpga_hls_ori*	*fpga_hls_opt*
0.58×10^{-5}	1.1×10^{-5}	0.1×10^{-5}

Table 6.2. *Running time comparison (s/pixel)*

Next, we can see that the optimized implementation can achieve the most potential running time gains of all of them. The contributors of this performance gain mainly include the target FPGA device and the optimization techniques. It is known that FPGAs have a high flexibility, which enables users to configure the architecture of the designs as they wish, while the CPU has a changeless architecture, which constrains the operation scheduling, i.e. von Neumann bottleneck and the thread number limit. Coupled with HLS, the former can benefit the performance of the design by providing a more efficient architecture and operation scheduling. According to our test, the final RTL optimized achieves around 5.29× and 10.4× speedups related to the two reference implementations, respectively.

6.4. Discussion and conclusion

In this chapter, we presented an embedded VHR satellite image segmentation design. First of all, FPGAs were selected as the target device for the desired design and a novel dedicated design framework was built using the HLS technique. This design flow can effectively accelerate the development cycles by facilitating algorithm analysis, design implementation and the exploration of design alternatives. Next, an active contour model and its LBM-based solver were prototyped and analyzed as the target image segmentation algorithm in a math-convenient environment. This algorithm has a high parallelism nature, which can greatly improve the design performance.

During the implementation process, the design is optimized depending on the features of HLS. In the evaluation experiments, it is seen that different implementations can produce high-quality image segmentation results in nature or disaster images taken by IKONOS or GeoEye-1. Compared to the two reference implementations on the CPU or FPGA, the optimized design has a 5.29–10.40× higher running time performance with a similar capacity in terms of maintainability. Therefore, it is concluded that the achievement has a great application prospect and potential in remote sensing-based natural disaster prevention and monitoring.

7

Real-time Image Processing with Very High-level Synthesis

Programming in a high abstraction level can facilitate the development of digital single processing systems. A high-level synthesis technique greatly benefits the R&D productivity of the FPGA-based designs and helps add to the maintainability of the products by automating the C-to-RTL conversion. However, due to the high complexity and computational intensity, the image processing designs usually necessitate a higher abstraction level than C-synthesis, and the current HLS tools do not have the ability of this kind.

This chapter briefly presents a very high-level synthesis (VHLS) method that allows fast prototyping and verifying the desired FPGA-based image processing in the Matlab environment. A heterogeneous design flow by using currently available tool kits is built in order to synthesize the algorithm behavior from the user level into the low register transfer level. The main objective is again to further reduce the complexity of design and give play to the advantages of FPGA related to the other devices.

7.1. VHLS motivation

High abstraction level design flows have significant value. Since the 1990s, many efforts have been made to develop a production quality HLS method for FPGA designs. However, all the currently available HLS tools are based on C-synthesis techniques, whereas a higher abstraction level is usually required due to the high complexity and computationally intensity of the target algorithms. We attempt to build an automatical VHLS framework with the following properties:

– to handle the algorithm behavior described in very high level languages, such as Matlab or OpenCL;

– to handle the code written without FPGA expertise or even not for FPGAs but the platforms of other types;

– to optimize the performances of the designs with the hardware constrains such as frequency or area of the target device;

– to automatically generate the desired RTL implementation in a short time rather than hours or even days;

– to be capable of being implemented by using the currently available electronic design automation (EDA) tools. This is important for industrial designs, because it can effectively reduce the R&D cost by helping to quickly build the desired design space exploration (DSE) framework and avoid the additional cost for the new tool kits.

To do this, we first select Matlab as the favorite user-level design environment for its advantages in terms of vector processing and powerful built-in image processing tools. Next, the challenges of Matlab-to-RTL synthesis are explored. In the exploration work, we find that Matlab is inherently a vector-based representation, whereas register-transfer languages are fully scalar-based. This issue may seriously constrain the benefits of the desired FPGA-available Matlab and is hardly solved by using a direct synthesis process. Therefore, we incorporate the source-to-source compile (SSC) in order to turn the nature of the source code from vector- into scalar-oriented programming. Finally, the generated code is classically synthesized via control and data flow extraction and RTL generation processes.

7.2. Image processing from Matlab to FPGA-RTL

Among all the programming languages favored by image processing designers, Matlab is widely used for the following reasons:

– it has a significant ability in matrix computing. Digital image processing is essentially matrix computation. The basic data element of Matlab is matric, and all the mathematical operations are designed to work on arrays or matrices. This can facilitate the programming tasks by freeing the users from the redundant works of data flow control or architecture specification;

– its library incorporates a rich database of built-in algorithms for image processing and computer vision applications, which can accelerate the algorithm tests without recompilation;

– it allows interaction between the users and their data, or even immediate statement running, which can help for file or variable tracking and simplifying prototyping or debugging tasks.

Unfortunately, since Matlab is inherently a vector-based representation, moving the designs from it into the register-transfer languages amenable to FPGAs is usually interrupted by the following challenges:

– operators in Matlab perform different operations depending on the type of the operands, whereas the functions of the operators in the register level languages are fixed;

– Matlab includes very simple and powerful vector operations such as the concatenation "[]" and column "x(:)" operators or "end" construct, which can be quite hard to map into RTL;

– Matlab supports "polymorphism", whereas register-transfer level programming does not. More precisely, functions in MATLAB are genetic and can process different types of input parameters. In the RTL behaviors, each parameter has only a single given type, which cannot change;

– Matlab supports dynamic loop bounds or vector size, whereas programming in RTL requires users to initialize them explicitly. Neither this initialization, nor any modification, can be done during the synthesis;

– the variables in Matlab can be reused for different contents (different types), whereas those in RTL cannot, as each variable has one unique type.

Considering that both Matlab and FPGAs greatly benefit the image processing applications, it is important to focus research efforts on the Matlab-based fast prototyping and verifying methodologies for FPGA image processing designs.

7.3. VHLS process presentation

Figure 7.1 displays the flowchart of this synthesis process, which is developed from the classical high-level synthesis technique. It consists of three steps: SSC, control and data extraction and RTL generation. For each step, interdependent tasks are executed. Since the two last steps are included in the classical HLS process, our presentation in this section focus on SSC.

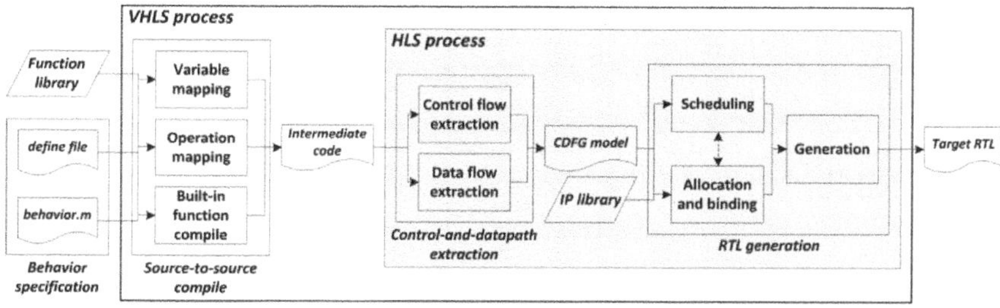

Figure 7.1. *Flowchart of the VHLS design*

As discussed in section 7.2, the nature of the vector-based representation of Matlab and its powerful built-in tools result in many challenges for synthesizing it to RTL. We summed up these issues as the following three problems: dynamic variable problem, operation polymorphism problem and built-in function problem, and solve them by compiling the Matlab source code into a second intermediate code.

7.3.1. *Dynamic variable*

Since the corresponding memories can be re-allocated over and over for the new contents with different lengths (types), Matlab users do not have to declare the types for variable initializing or allocate the right amount of memory for vector variable before each use. The variables can automatically change their lengths (types) or dimensions depending on the content to be held. However, none of the currently available register transfer languages have the dynamic variable ability like this. Therefore, as shown in Figure 7.2(a), we must explicitly allocate enough storage for every variable: a manually specified *define file* is required for memory allocation. Furthermore, when a variable is dynamically reused in the source code, such as x and X in the example, some additional definitions with other names are necessitated, i.e. *x_1* and *X_1* in the example.

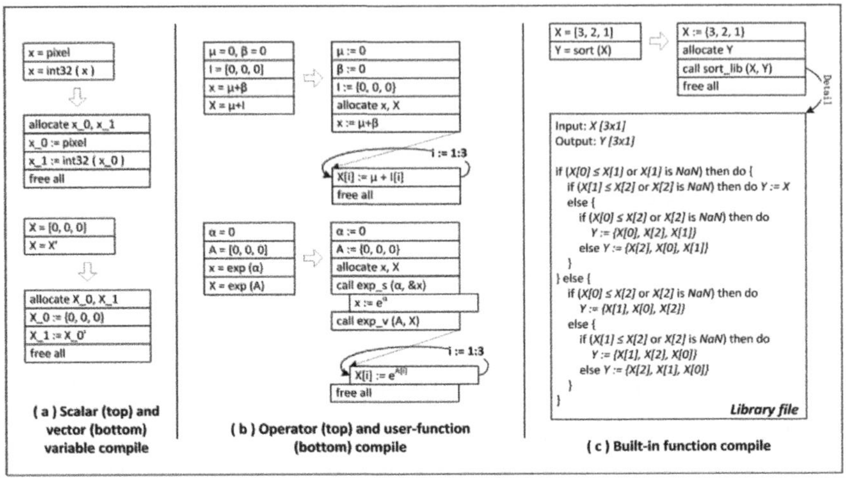

Figure 7.2. *Source-to-source compilation strategies for VHLS*

7.3.2. *Operation polymorphism problem*

Matlab allows operator and function polymorphism. That is, its operators or functions may support either a matrix or a scalar as a second operand or argument, and their returns are also changed depending on the inputs. For the operators, the nature of each invocation has to be determined first. If scalar, they are mapped directly as the corresponding monomorphism ones of the target languages, else they are replaced with a loop construct as shown in the top of Figure 7.2(b). For the function mapping, the types of all the arguments should be determined as well. Depending on the invocation natures, we must create multiple different versions for the same Matlab functions to satisfy the operations of different invocations. As shown in the bottom of Figure 7.2(b), two exponential computation functions, *exp_s()*, and *exp_v()*, are created to handle the scalar and vector invocations, respectively, of *exp()*.

We can see that either operator or function mapping may necessitate additional loop constructs in the present or deeper function level. In RTL, all the loop boundaries must be initialized in advance instead of using the unknown variables due to the fact that FPGAs support only static compiling. Here, we use the vector information, which has been explicitly defined in *define file*, to help to compute the boundaries of the loops generated for operation polymorphism problem. Meanwhile, it should be noted that different vector dimensions may require more additional function versions even for the invocations that have same nature.

7.3.3. *Built-in function problem*

The built-in algorithms/functions of Matlab provide users many benefits of facilitating their algorithm specification, but they are usually invisible to a third-party compiler and detected as undefined functions when invoked. Therefore, we need to build a new library by using the synthesizable code for these powerful algorithms. As shown in Figure 7.2(c), the *sort* function of Matlab is respecified in the library. When it is invoked in the source code, the corresponding routine can easily be compiled by including its specification in the generated file. Since the built-in functions of Matlab also have the polymorphism nature, we create different versions of each. Depending on the argument types, the right version is re-targeted to the invocation in the source code.

7.4. VHLS implementation issues

According to section 7.3, we can see that the VHLS method is difficult to implement. In this section, we will discuss some VHLS implementation issues.

7.4.1. *Work flow*

The work flow for the VHLS method is shown in Figure 7.3, within which the following tasks should be done:

1) users first specify and functionally verify their algorithm behavior by using classical Matlab;

2) next, the verified Matlab code is compiled into the intermediate code. Due to the compatibility problem between the tools, the code needs to be further transformed for the subsequent steps. Additionally, a second function verification can be done by using the corresponding compilers if necessary;

3) third, the verified intermediate code is synthesized into RTL through a synthesis tool with the user-specified and hardware constraints;

4) fourth, the generated RTL is evaluated. If the results satisfy the design requirement, go to the next step; otherwise go back to the first step;

5) finally, the evaluated RTL is send back to the Matlab environment for cosimulation. If its logic verification is satisfied, work end, otherwise go back to the first step.

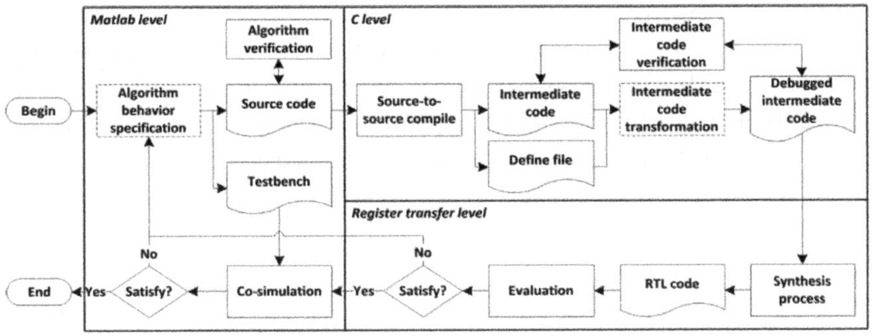

Figure 7.3. *VHLS-based work flow*

7.4.2. *Intermediate code versus RTL*

The first step is to compile the source code from Matlab into an intermediate code. ANSI C is selected for the following reasons:

– it is a scalar-based language with high compatibility, which can satisfy most requirements of the synthesis process related to language natures;

– it can be compiled as either source or destination code by many mature source-to-source compilers, which can facilitate the discovery of the alternative solutions for SSC task. This is important for DSE because the description methods or code style could influence the performance of the generated RTL [CON 13];

– it is supported by most available FPGA synthesis tools. Since the qualities of the generated RTL are essentially decided by the synthesis strategies, i.e. scheduling and binding algorithms, so different tools may result in different design results. Being supported by multiple synthesis tools allows easy experiments for solution improvements.

Based on this decision, Matlab Coder is selected for the Matlab-to-C conversion. Meanwhile, for the C synthesis tool, we select Auto Pilot.

7.4.3. *SSC versus HLS*

Since Matlab Coder is not specially designed for Auto Pilot, the C code generated by it cannot be directly used for synthesis. This issue is caused by: (1) Matlab Coder handles memory allocations by using *malloc*, whereas this C library function is not supported by Auto Pilot, and (2) static variable is not amenable to Auto Pilot, because FPGAs do not have the storage element of this type. Therefore, before Control and Data Flow Graph extraction, some C-to-C code transformations must be done. We map the memory allocation statements in intermediate C code to the common array declarations, while the static variable is mapped into a variable out of the top entity and stored in the external memory.

7.4.4. *Verification and evaluation*

According to Figure 7.3, we can see that three verification and one evaluation tools are required. The source code can be easily verified by using the compiler of Matlab. Meanwhile, we select Intel C++ Compiler to debug the intermediate C code and evaluate the generated RTL by using Auto Pilot directly. Since the algorithm behavior is specified within a Matlab environment, the final cosimulation should be performed between Matlab and RTL. We can address this problem by using System Generator/Simulink [XIL 12a]. This tool provides a visual programming environment that can facilitate the building of the testbench and profit from the data base already configured in Matlab.

7.5. Future work for real-time image processing with VHLS

This chapter proposes a VHLS method for image processing designs by combining the recent source-to-source compilation and HLS techniques. It provides a very high abstraction level development environment to the users, which can greatly benefit the development productivity by automating the Matlab-to-RTL synthesis process. This synthesis method has been implemented by using currently available EDA tools and its feasibility is verified. Furthermore, the VHLS will not increase the complexity of the algorithm behaviors described using Matlab in routine level, even if they are not specially developed for FPGAs. Therefore, this method can effectively facilitate the transplantation between the different platforms, such as CPU versus FPGA.

On the other hand, some new issues and challenges are found as well. The first is that manual code transformation is still needed due to the incompatibility between the selected source-to-source compiler and HLS tool. However, this issue should not be considered hard to fix, because Matlab-to-C conversion is quite a mature technique in software engineering, taking the constraints of HLS-available C into the compiling process can be easily realized through the currently available techniques. The second one is that, in HLS, there still exist many potential optimization opportunities that may further benefit the performances of the generated RTL, especially the running speed, but they are not yet discussed. Consequently, some SSC-based optimization strategies may be developed to improve the quality of the generated RTL.

Bibliography

[ACK 99] ACKENHUSEN J.G., *Real Time Signal Processing: Design and Implementation of Signal Processing Systems*, Prentice Hall PTR, Indianapolis, 1999.

[ADS 16] ADSC, "ADSC research highlights: synthesize hardware, without hardware expertise", available at: https://adsc.illinois.edu/research/adsc-research-highlights/adsc-research-highlights-synthesize-hardware-without-hardware-expe, January 2016.

[AGU 05] AGUIRRE M.A., TOMBS J.N. *et al.*, "Microprocessor and FPGA interfaces for in–system co–debugging in field programmable hybrid systems", *Microprocessors and Microsystems*, vol. 29, pp. 75–85, 2005.

[AKG 14] AKGÜN D., "A practical parallel implementation for tdlms image filter on multicore processor", *Journal of Real–Time Image Processing*, pp. 1–12, 2014.

[ALT 16] ALTERA, "NIOS II Gen2 Processor Reference Guide", available at: https://www.altera.com/documentation/iga1420498949526.html, 2016.

[AYG 09] AYGUADE E., COPTY N., DURAN A. *et al.*, "The design of openmp tasks", *IEEE Transactions on Parallel and Distributed Systems*, vol. 20, no. 3, pp. 404–418, 2009.

[BAR 02] BARAT F., LAUWEREINS R., DECONINCK G. *et al.*, "Reconfigurable instruction set processors from a hardware/software perspective", *IEEE Transactions on Software Engineering*, vol. 28, no. 9, pp. 847–862, 2002.

[BAR 17] BARNEY B., "Posix threads programming", available at: https://computing.llnl.gov/tutorials/pthreads/, 2017.

[BAR 14] BARTOVSKY J., DOKLADAL P., DOKL E. *et al.*, "Parallel implementation of sequential morphological filters", *Journal of Real-Time Image Processing*, vol. 9, no. 2, pp. 315–327, 2014.

[BER 10] BERGAN T., ANDERSON O., "CoreDet: A compiler and runtime system for deterministic multithread execution", *ACM Sigplan Notices*, vol. 45, no. 3, pp. 53–64, 2010.

[BIS 13] BISWAL P., MONDAL P., BANERJEE S. *et al.*, "Parallel architecture for accelerating affine transform in high–speed imaging systems", *Journal of Real–Time Image Processing*, vol. 8, no. 1, pp. 69–79, 2013.

[BIS 06] BISWAS P., BANERJEE S., "ISEGEN: An intarative improvement-based ISE generation technique for fast customization of processors", *IEEE Transactions on VLSI Systems*, vol. 14, no. 7, pp. 754–762, 2006.

[BRO 07] BROST V., YANG F., PAINDAVOINE M. *et al.*, "Multiple modular VLIW processors based on FPGA", *Journal of Electronic Imaging*, SPIE, vol. 16, no. 2, pp. 1–10, 2007.

[CAD 11] CADENCE, "C–to–Silicon Compiler High-Level Synthesis", Cadence Design Systems, Inc, available at: https://www.cadence.com/rl/Resources/datasheets/C2Silicon_ds.pdf, 2011.

[CHA 89] CHAM W.K., "Development of integer cosine transforms by the principle of dyadic symmetry", *IEEE Proceeding I: Communications, Speech and Vision*, vol. 136, no. 4, pp. 276–282, 1989.

[CHE 77] CHEN C.S., FRALICK S.A., "A fast computational algorithm for the discrete cosine transform", *IEEE Transactions on Communication*, vol. 25, pp. 1004–1009, 1977.

[CHE 07] CHEN X., MASKELL D.L., SUN Y. *et al.*, "Fast identification of custom instructions for extensible processors", *IEEE Transactions on Computer-Aided Design of Integrated Circuits and Systems*, vol. 26, no. 2, pp. 359–368, 2007.

[CHE 08] CHE S., LI J., SHEAER J. *et al.*, "Accelerating compute-intensive applications with gpus and fpgas", *Symposium on Application Specific Processors (SASP)*, pp. 101–107, Anaheim, California, USA, June 2008.

[CHI 94] CHIODO M., GIUSTO P.A., JURECSKA A. *et al.*, "Hardware-software co-design of embedded systems", *Micro,* IEEE, vol. 14, no. 4, pp. 26–36, 1994.

[CHO 10] CHOI S., "Fast and robust extraction of optical and morphological properties of human skin using a hybrid stochastic-deterministic algorithm: Monte–Carlo simulation study", *Lasers in Medical Sciences*, vol. 25, no. 5, pp. 733–741, 2010.

[CON 11] CONG J., LIU B., NEUENDORER S. *et al.*, "High-level synthesis for fpgas: From prototyping to deployment", *IEEE Transactions on Computer-Aided Design of Integrated Circuits and Systems*, vol. 30, no. 4, pp. 473–491, 2011.

[DEM 11] DEMING C., ERIC L., KYLE R. *et al.*, Hardware synthesis without hardware expertise, Technical Report, Advanced Digital Sciences Center (ADSC), University of Illinois, 2011.

[GEO 10] GEOFFRAY N., THOMAS G., "VMKit: a substrate for managed runtime environments", *ACM Sigplan Notices*, vol. 45, no. 7, pp. 51–61, 2010.

[GLE 04] GLESNER M., HOLLSTTEIN T. *et al.*, "Reconfigurable platforms for ubiquitous computing", *Proceedings of the First Conference on Computer Frontiers*, pp. 377–389, New York, USA, 2004.

[HAS 11] HASAN M.Z., SOTIRIOS S.G., "Customized kernel execution on reconfigurable hardware for embedded applications", *Journal of Microprocessors and Microsystems*, vol. 33, pp. 211–220, 2011.

[HOS 13] HOSSEINI F., FIJANY A., SAFARI S. *et al.*, "Fast implementation of dense stereo vision algorithms on a highly parallel simd architecture", *Journal of Real–Time Image Processing*, vol. 8, no. 4, pp. 421–435, 2013.

[HUA 10] HUANG R., DENG D.Y., SUH G.E. *et al.*, "Orthrus: efficient software integrity protection on multi-cores", *ACM Sigplan Notices*, vol. 45, no. 3, pp. 371–383, 2010.

[IEE 06] IEEE Std., IEEE standard for Verilog hardware description language, IEEE Std 1364–2001, IEEE, 2006.

[IEE 11] IEEE Std., Behavioral languages - part 1-1: *VHDL* language reference manual, IEEE Std 1076–2008, IEEE, 2011.

[INT 08] Intel, "Intel C++ Compiler User and Reference Guides", 304968–022us edition, available at: http://www.physics.udel.edu/~bnikolic/QTTG/shared/docs/intel_c_user_and_reference_guide.pdf, 2008.

[JAC 08] JACQUES S.L., Spectroscopic determination of tissue optical properties using optical fiber spectrometer, Technical Report, available at: http://omlc.ogi.edu/news/apr08/skinspectra/index.html, 2008.

[JOL 11] JOLIVOT R., Development of an imaging system dedicated to the acquisition, analysis and multispectral characterisation of skin lesions, PhD Thesis, University of Burgundy, 2011.

[JOL 13] JOLIVOT R., BENEZETH Y., MARZANI F. *et al.*, "Skin parameter map retrieval from a dedicated multispectral imaging system applied to dermatology/cosmetology", *International Journal of Biomedical Imaging*, vol. 15, p. 15, 2013.

[JUN 11] JUNG Y.K., "Hardware/software co–reconfigurable instruction decoder for adaptive multi–core DSP architectures", *Journal Signal Processing System*, vol. 62, pp. 273–285, 2011.

[KAR 08] KARURI K., CHATTOPADHYAY A., "A design flow for architecture exploration and implementation of paritially reconfigurable processors", *IEEE Transactions on VLSI systems*, vol. 16, no. 10, pp. 1281–1294, 2008.

[KEH 06] KEHTARNAVAZ N., GAMADIA M., *Real–Time Image and Video Processing: from Research to Reality*, Morgan & Claypool Publishers, San Rafael, 2006.

[KEH 11] KEHTARNAVAZ N., *Real–Time Digital Signal Processing Based on the TMS320C6000*, Elsevier, Amsterdam, 2011.

[KES 10] KESTUR S., DAVIS J.O., WILLIAMS O. *et al.*, "Blas comparison on FPGA, CPU and GPU", *VLSI (ISVLSI), IEEE Computer Society Annual Symposium*, pp. 288–293, Lixouri, Greece, 2010.

[KUB 31] KUBELKA P., MUNK F., "Ein beitrag zur optik der farbanstriche", *Zeitschrift für technische Physik*, vol. 9, pp. 593–601, 1931.

[KYR 12] KYRKOU C., THEOCHARIDES T., "A parallel hardware architecture for real-time object detection with support vector machines", *IEEE Transactions on Computers*, vol. 61, no. 6, pp. 831–842, 2012.

[LAM 09] LAM S.K., SRIKANTHAN T., "Rapid design of area–efficient custom instructions for reconfigurable embedded processing", *Journal of System Architecture*, vol. 55, pp. 1–14, 2009.

[LEE 97] LEE Y., "A cost effective architecture for 8x8 two–dimensional dct/idct using direct method", *IEEE Transactions on Circuits and Systems for Video Technology*, vol. 7. no. 3, pp. 459–467, 1997.

[LIA 12] LIANG Y., RUPNOW K., LI Y. *et al.*, "High-level synthesis: Productivity, performance, and software constraints", *Journal of Electrical and Computer Engineering*, vol. 14, p. 14, 2012.

[LLV 17] LLVM., "The LLVM Compiler Infrastructure", available at: http://llvm.org, 2017.

[LOE 89] LOEFFLER C., LIGHTENBERG A, "Practical fast 1–D dct algorithms with 11 multiplications", *Proceeding of IEEE ICASSP'89*, vol. 2, pp. 988–991, New York, USA, 1989.

[LOO 02] LOO S.M, WELLS B., "Handel-C for rapid prototyping of VLSI coprocessors for real-time systems", *Proceedings of the 34th Southeasterm Symposium System Theory*, pp. 6–10, Huntsville, Alabama, USA, 2002.

[LU 08] LU H., FORIN A., "Automatic processor customization for zero–overhead online software verification", *IEEE Transactions on VLSI systems*, vol. 16, no. 10, pp. 1346–1357, 2008.

[MA 07] MA S., KUO C.-C.J., "High-definition video coding with super-macroblocks", *Visual Communications and Image Processing*, SPIE Proceedings, vol. 6508, 2007.

[MAT 15] MATHWORKS, "Getting Started with MATLAB", available at: https://www.mathworks.com/help/pdf_doc/matlab/getstart.pdfr, 2015.

[MEE 12] MEEUS W., VAN BEECK K., GOEDEME T. *et al.*, "An overview of today's high-level synthesis tools", *Design Automation for Embedded Systems*, vol. 16, no. 3, pp. 31–51, 2012.

[MOR 12] MORELAND T.B., *Introduction to Video and Image Processing, Building Real Systems and Applications*, Springer, Berlin Heidelberg, 2012.

[MYE 09] MYERS D.G., "Image processing", *Electrical Engineering*, vol. 1, pp. 397–433, 2009.

[NAT 07] NATIONAL INSTRUMENTS, "Getting Started with LabVIEW", available at: http//itech.fgcu.edu/faculty/zalewski/cda4170/files/LV_Getting_Started.pdf, 2007.

[NVI 11] NVIDIA, "Intro to OpenCL", available at: http://www.cc.gatech.edu/~vetter/keeneland/tutorial-2011-04-14/06-intro_to_opencl.pdf, 2011.

[NVI 12] NVIDIA, "Information to CUDA C", available at: http://www.cc.gatech.edu/~vetter/keeneland/tutorial-2012-02-20/07-intro_to_cuda_c.pdf, 2012.

[PAN 01] PANDA P.R., "SYSTEM C – a modeling platform supporting multiple design abstractions", *Proceedings of the 14th International Symposium System Synthesis*, pp. 75–80, Montreal, Quebec, Canada, August 2001.

[PRO 14] PROST–BOUCLE A., MULLER O., ROUSSEAU F., "Fast and standalone design space exploration for high-level synthesis under resource constraints", *Journal of Systems Architecture*, vol. 60, no. 1, pp. 79–93, 2014.

[RUP 11] RUPNOW K., LIANG Y., LI Y. *et al.*, "High level synthesis of stereo matching: Productivity, performance, and software constraints", *IEEE International Conference on Field–Programmable Technology*, pp. 1–8, New Delhi, India, 2011.

[SAM 10] SAMSUNG ELECTRONICS CO. LTD and BRITISH BROADCASTING CORPORATION, "HEVC Reference software HM 5.0", available at: https://hevc.hhi.fraunhofer.de/svn/svn_HEVC Software/tags/HM-5.2/, 2010.

[SET 10] SETO K., FUJITA M., "Custom Instruction generation for configurable processors with limited numbers of operands", *IPSJ Transactions on System LSI Design Methodology*, vol. 3, pp. 57–68, 2010.

[SHI 08] SHIN D., GERSTLAUER A., "An interactive design environment for C–based high–level synthesis of RTL processors", *IEEE Transactions on VLSI systems*, vol. 16, no. 4, pp. 466–475, 2008.

[STR 08] STRATTON J.A., STONE S.S., HWU W.-M. *et al.*, "An efficient implementation of cuda kernels for multi-core cpus", in NELSON AMARAL J. (ed.), *Languages and Compilers for Parallel Computing*, Springer–Verlag, Berlin, 2008.

[SYN 10] SYNOPSYS, "Synphony C Compiler", available at: http://www.scanru.ru/file_link.php?fid=831, 2010.

[SYS 89] SYSWERDA G., "Uniform crossover in genetic algorithms", *Proceedings of the 3rd International Conference on Genetic Algorithms*, pp. 2–9, San Francisco, CA, USA, 1989.

[TAK 79] TAKATANI S., GRAHAM M.D., "Theoretical analysis of diffuse reflectance from a two–layer tissue model", *IEEE Transactions on Biomedical Engineering*, vol. 26, no. 12, pp. 656–664, 1979.

[TIE 05] TIENSYRJA K., CUPAK M., SYSTEM C. *et al.*, "Based system-level design for reconfigurable System-on-Chip", in BOULET P. (ed.), *Advances in Design and Specification Languages for SoCs*, Springer, Berlin, 2005.

[TOL 09] TOL M.V., JESSHOPE C., LANKAMP M. *et al.*, "An implementation of the fSANEg virtual processor using fPOSIXg threads", *Journal of Systems Architecture*, vol. 55, no. 3, pp. 162–169, 2009.

[UGU 10] UGUR K., ANDERSSON K., "High performance, low complexity video coding and the emerging HEVC standard", *IEEE Transactions on Circuits and Systems for Video Technology*, vol. 20, no. 12, pp. 1688–1697, 2010.

[VIA 05] VIATOR J.A., JUNG B., SVAASAND L.O. *et al.*, "Determination of human skin optical properties from spectrophotometric measurements based on optimization by genetic algorithms", *Journal of Biomedical Optics*, vol. 10, no. 2, 2005.

[VIL 10] VILLARREAL J., PARK A., NAJJAR W. *et al.*, "Designing modular hardware accelerators in c with roccc 2.0", *18th IEEE Annual International Symposium on Field-Programmable Custom Computing Machines*, pp. 127–134, Charlotte, North Carolina, USA, 2010.

[WAK 05] WAKABAYASHI K., "Cyberworkbench: integrated design environment based on c–based behavior synthesis and verification", *IEEE VLSI-TSA International Symposium on Design, Automation and Test*, pp. 173–176, Hsinchu, Taiwan, 2005.

[WAN 10] WANG G., CATAPULT C., Synthesis Work Flow Tutorial, ELEC 522 Advanced VLSI Design, Rice University, version 1.3, October 2010.

[XIL 09a] XILINX, "Micro Blaze Soft Processor Core", available at: http://www.xilinx.com/products/design_resources/proc_central/microblaze_faq.pdf, 2009.

[XIL 09b] XILINX, "ISE design suite 11", available at: http://www.xilinx.com/tools/designtools.htm, 2009.

[XIL 09c] XILINX, VIRTEX–6, "ds150 virtex-6 family overview", available at: http://www.xilinx.com/support/documentation/data_sheets, 2009.

[XIL 12a] XILINX, "System Generator for DSP – Getting Started Guide", available at: http://www.xilinx.com/support/documentation/sw_manuals/xilinx14_5/sysgen_gs.pdf, 2012.

[XIL 12b] XILINX, "Vivado Design Suite User Guide", ug902(2012.2) edition, Xilinx, July 2012.

[XIL 13a] XILINX, Introduction to fpga design with vivado high–level synthesis, Technical Report UG998 (v1.0), 2013.

[ZHA 08] ZHANG Z., FAN Y., JIANG W. *et al.*, "Autopilot: A platform–based ESL synthesis system", in COUSSY P., MORAWIEC A. (eds), *High–Level Synthesis*, Springer, Netherlands, 2008.

[ZOU 12] ZOU D., DOU Y., XIA F. *et al.*, "Optimization schemes and performance evaluation of smith-waterman algorithm on CPU, GPU and FPGA", *Concurrency and Computation: Practice and Experience*, vol. 24, no. 14, pp. 1625–1644, 2012.

Index

O, P, R

S, V

Other titles from

ISTE

in

Digital Signal and Image Processing

2017

CESCHI Roger, GAUTIER Jean-Luc
Fourier Analysis

CHARBIT Maurice
Digital Signal Processing with Python Programming

FEMMAM Smain
Fundamentals of Signals and Control Systems
Signals and Control Systems: Application for Home Health Monitoring

MAÎTRE Henri
From Photon to Pixel – 2nd edition

PROVENZI Edoardo
Computational Color Science: Variational Retinex-like Methods

2015

BLANCHET Gérard, CHARBIT Maurice
Digital Signal and Image Processing using MATLAB®
Volume 2 – Advances and Applications:The Deterministic Case – 2nd edition
Volume 3 – Advances and Applications: The Stochastic Case – 2nd edition

CLARYSSE Patrick, FRIBOULET Denis
Multi-modality Cardiac Imaging

GIOVANNELLI Jean-François, IDIER Jérôme
Regularization and Bayesian Methods for Inverse Problems in Signal and Image Processing

2014

AUGER François
Signal Processing with Free Software: Practical Experiments

BLANCHET Gérard, CHARBIT Maurice
Digital Signal and Image Processing using MATLAB®
Volume 1 – Fundamentals – 2nd edition

DUBUISSON Séverine
Tracking with Particle Filter for High-dimensional Observation and State Spaces

ELL Todd A., LE BIHAN Nicolas, SANGWINE Stephen J.
Quaternion Fourier Transforms for Signal and Image Processing

FANET Hervé
Medical Imaging Based on Magnetic Fields and Ultrasounds

MOUKADEM Ali, OULD Abdeslam Djaffar, DIETERLEN Alain
Time-Frequency Domain for Segmentation and Classification of Non-stationary Signals: The Stockwell Transform Applied on Bio-signals and Electric Signals

NDAGIJIMANA Fabien
Signal Integrity: From High Speed to Radiofrequency Applications

PINOLI Jean-Charles
Mathematical Foundations of Image Processing and Analysis
Volumes 1 and 2

TUPIN Florence, INGLADA Jordi, NICOLAS Jean-Marie
Remote Sensing Imagery

VLADEANU Calin, EL ASSAD Safwan
Nonlinear Digital Encoders for Data Communications

2013

GOVAERT Gérard, NADIF Mohamed
Co-Clustering

DAROLLES Serge, DUVAUT Patrick, JAY Emmanuelle
Multi-factor Models and Signal Processing Techniques: Application to Quantitative Finance

LUCAS Laurent, LOSCOS Céline, REMION Yannick
3D Video: From Capture to Diffusion

MOREAU Eric, ADALI Tulay
Blind Identification and Separation of Complex-valued Signals

PERRIN Vincent
MRI Techniques

WAGNER Kevin, DOROSLOVACKI Milos
Proportionate-type Normalized Least Mean Square Algorithms

FERNANDEZ Christine, MACAIRE Ludovic, ROBERT-INACIO Frédérique
Digital Color Imaging

FERNANDEZ Christine, MACAIRE Ludovic, ROBERT-INACIO Frédérique
Digital Color: Acquisition, Perception, Coding and Rendering

NAIT-ALI Amine, FOURNIER Régis
Signal and Image Processing for Biometrics

OUAHABI Abdeljalil
Signal and Image Multiresolution Analysis

2011

CASTANIÉ Francis
Digital Spectral Analysis: Parametric, Non-parametric and Advanced Methods

DESCOMBES Xavier
Stochastic Geometry for Image Analysis

FANET Hervé
Photon-based Medical Imagery

MOREAU Nicolas
Tools for Signal Compression

2010

NAJMAN Laurent, TALBOT Hugues
Mathematical Morphology

2009

BERTEIN Jean-Claude, CESCHI Roger
Discrete Stochastic Processes and Optimal Filtering – 2nd edition

CHANUSSOT Jocelyn *et al.*
Multivariate Image Processing

DHOME Michel
Visual Perception through Video Imagery

GOVAERT Gérard
Data Analysis

GRANGEAT Pierre
Tomography

MOHAMAD-DJAFARI Ali
Inverse Problems in Vision and 3D Tomography

SIARRY Patrick
Optimization in Signal and Image Processing

2008

ABRY Patrice *et al.*
Scaling, Fractals and Wavelets

GARELLO René
Two-dimensional Signal Analysis

HLAWATSCH Franz *et al.*
Time-Frequency Analysis

IDIER Jérôme
Bayesian Approach to Inverse Problems

MAÎTRE Henri
Processing of Synthetic Aperture Radar (SAR) Images

MAÎTRE Henri
Image Processing

NAIT-ALI Amine, CAVARO-MENARD Christine
Compression of Biomedical Images and Signals

NAJIM Mohamed
Modeling, Estimation and Optimal Filtration in Signal Processing

QUINQUIS André
Digital Signal Processing Using Matlab

2007

BLOCH Isabelle
Information Fusion in Signal and Image Processing

GLAVIEUX Alain
Channel Coding in Communication Networks

OPPENHEIM Georges *et al.*
Wavelets and their Applications

2006

CASTANIÉ Francis
Spectral Analysis

NAJIM Mohamed
Digital Filters Design for Signal and Image Processing

Printed and bound by CPI Group (UK) Ltd, Croydon, CR0 4YY
15/09/2026
14962342-0004